MASTERPIECES

OF THE MINERAL WORLD

WENDELL E. WILSON
Editor, The Mineralogical Record

JOEL A. BARTSCH
President
Houston Museum of Natural Science

with
MARK MAUTHNER
Associate Curator
Houston Museum of Natural Science

MASTERPIECES

OF THE MINERAL WORLD

TREASURES FROM THE HOUSTON MUSEUM OF NATURAL SCIENCE

Published by The Houston Museum of Natural Science and
The Mineralogical Record *in association with Harry N. Abrams, Inc., Publishers*

TABLE OF CONTENTS

Library of Congress control number: 2004110408
Edited by John S. White, Amy Teschner, and Paul G. Bernhard
Conceived, designed, and produced by Studio Blue, Chicago, www.studioblueinc.com
Color separations by Prographics, Rockford, IL.
Printed in U.S.A.
10 9 8 7 6 5 4 3 2 1

Distributed by
Harry N. Abrams, Inc.
100 Fifth Avenue
New York, N.Y. 10011
www.abramsbooks.com

Abrams is a subsidiary of
LA MARTINIÈRE GROUPE

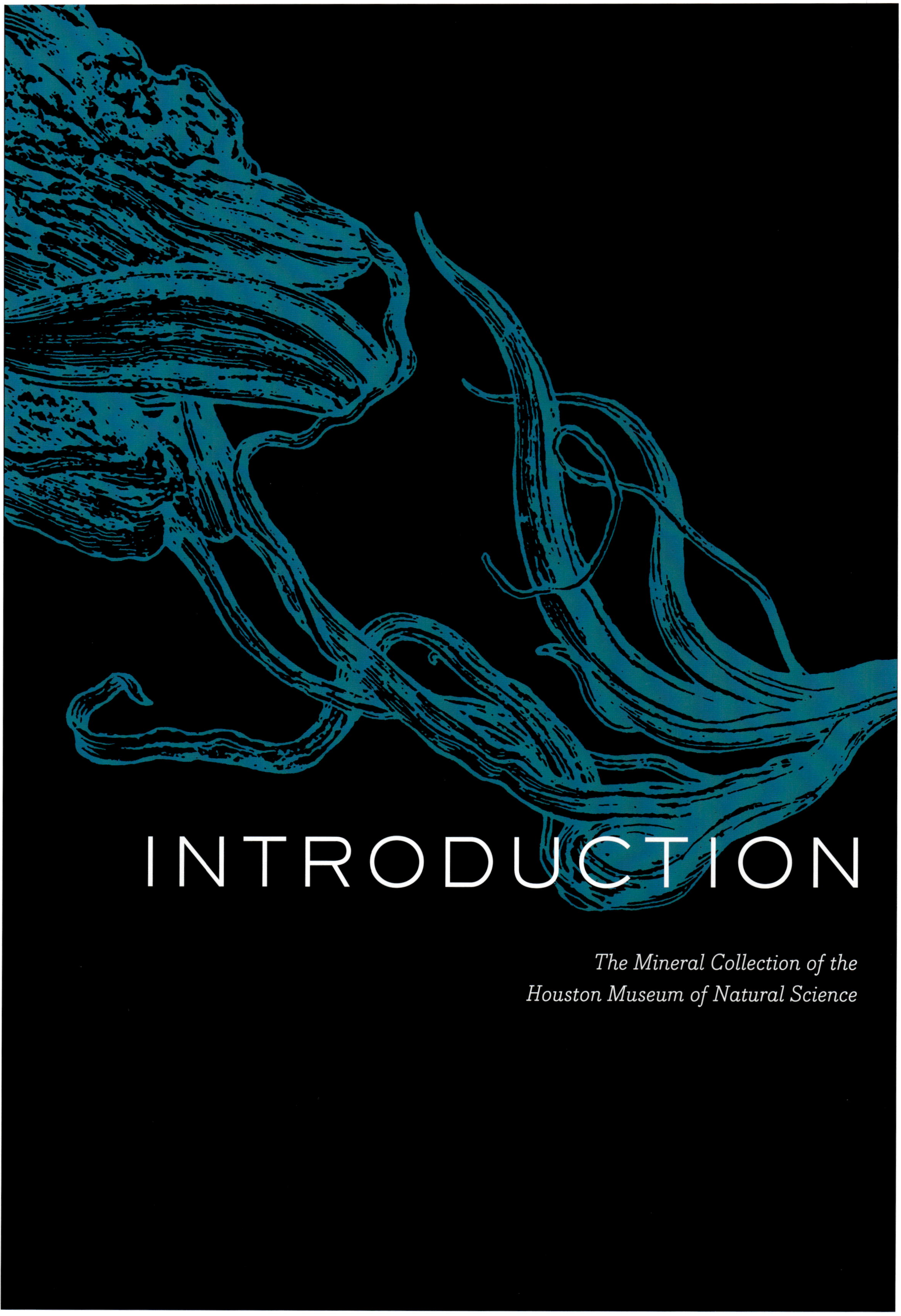

INTRODUCTION

The Mineral Collection of the Houston Museum of Natural Science

The great repositories of minerals today are the major public museums and institutions around the world that, through the centuries, have gathered together and preserved more than five million mineral specimens. Many of these museums were founded on pre-existing private collections, and over the past century they have continued to benefit from the acquisition of important collections assembled by individuals.

Today there are more private collections being built, some of which will eventually be absorbed by other museums. But there are also tens of thousands of small collections owned by people who do not aspire to build museum display collections, but are rather pursuing the immense satisfaction and enjoyment that comes from developing a personal collection. The Houston Museum of Natural Science, built on significant private collections and donations, exemplifies what may be the single factor that distinguishes 21st-century collecting from that of previous centuries: its relatively widespread popularity. There has probably never been a time in history when so many people were so involved with collecting minerals.

The Houston Museum of Natural Science has acquired several collections of worldwide mineral specimens. Major John Milsaps built one such collection. Acquired in 1927, the Milsaps Collection, along with other collections and donations, formed the core of the museum's gem and mineral collection. The best, however, was yet to come.

During the mid-1970s there arrived on the mineral collecting scene a determined Texas oilman and his wife who set out to assemble one of the world's finest mineral collections. Ann and Perkins Sams teamed with Paul Desautels, former curator of gems and minerals at the U.S. National Museum of Natural History (Smithsonian Institution) to accomplish their goal. The latter part of the 1970s and the first half of the 1980s saw frenzied activity in the mineral market, in large part due to the collecting activities of this illustrious trio.

An integral part of their collecting strategy was to acquire entire collections from other major collectors, thereby ensuring that they would obtain a significant number of the finest pieces available in as short a time as possible. Several important collections were absorbed, including the collection of San Diego County, California, minerals assembled by David P. Wilber; Miriam and Julius Zweibel's superb worldwide collection featuring some of the finest pieces ever recovered from the Tsumeb mine in Namibia; the worldwide tourmaline collection of Charles Key; the collection of Paul Obeniche of France; and some superb specimens from the collection of Ed Swoboda – to name just a few.

As the collection grew it became apparent that their approach to collecting was anything but subtle. For example, rather than selecting one or two specimens of azurite from a major find at the Tsumeb mine in Namibia, Sams and Desautels bought the entire pocket! Determined to assemble a major suite of tourmalines, they purchased several hundred over the course of just a few months. As a result of this aggressive collecting, they were able to capture some of the finest known specimens of numerous minerals. Juxtaposed with these stunning pieces are some of the world's finest crystals of some exceedingly rare minerals. This combination – pieces of unrivaled beauty complemented by specimens of extreme rarity – broadens the appeal of the collection to the layperson and connoisseur alike.

In a few short years Ann and Perkins Sams had indeed assembled one of the finest private mineral collections in the world, and they soon began looking for a home for their magnificent minerals, with the hope that the collection would remain in Texas. In 1984, Ernest H. Cockrell, president of the board of trustees of the Houston Museum of Natural Science, became aware that the Sams Collection might be available. Cockrell, a collector himself, was able to appreciate the importance of the collection. He recognized an unusual opportunity for the museum to acquire a collection that, in fact, consisted of several of the world's most important private collections, constituting a cornerstone for future acquisitions. With Cockrell's guidance, and the very generous support of the citizens of Houston, especially the Cullen and Brown foundations, the board of trustees made the commitment to acquire the collection, thereby establishing the Houston Museum of Natural Science as one of the world's major repositories for fine mineral specimens.

Over the last decade, the museum has continued to benefit from the acquisition of individual pieces as well as from the acquisition of other major collections, both in part and in whole. The museum has also gained immeasurably from some extraordinarily generous anonymous donations and bequests.

As the new millennium unfolds, there is no end in sight to the collector's passion for acquiring new finds – or for the ambitious museum to find a means to place them on public display. A few of the more important collections acquired by the Houston Museum of Natural Science over the last several years are briefly described here.

Dr. Edward E. and Ann David (1925–)

A former White House science advisor and president of Exxon Research, Dr. David developed a passion for minerals that led him to assemble one of the 20th century's preeminent collections, featuring a broad selection of worldwide specimens of superior aesthetic quality. The core of this collection was acquired by the Houston Museum of Natural Science in 1995, placing our gem and mineral collection among the world's finest.

F. John Barlow (1928–2004)

Wisconsin industrialist John Barlow put together another of the world's finest private mineral collections over a span of more than 30 years. Barlow's more than 5,000 pieces included the definitive suite of Wisconsin minerals, many of the world's finest gold crystals, and numerous legendary classics, including an exceptional tanzanite crystal (plate 79) from Merelani, Tanzania. A portion of the Barlow Collection found a new home at the Houston Museum of Natural Science in 1999.

The Duval Collection (1968–2002)

Assembled under the direction of S. H. Martin, general manager of Duval Mining Corporation's Battle Mountain (Nevada) and Mineral Park (Arizona) mineral properties, the Duval Collection (a gift from Pennzoil-Quaker State Company with the support of Shell Oil Products US), includes dazzling specimens of native copper, azurite, amethyst, selenite, and many others. Housed at the top of the Pennzoil Building for more than 20 years, this magnificent collection consists of numerous breathtaking pieces from around the world, all chosen for their beauty of form and color. Highlights of the collection include one of the world's largest "gem quality" turquoise nuggets and a magnificent cluster of transparent quartz crystals that weighs almost 2,000 pounds!

It is certainly fair to say that the mineral collection at the Houston Museum of Natural Science represents considerable expenditures of time and money. But more important, as a collection of collections it ultimately reflects decades, if not centuries, of connoisseurship in minerals. For while some specimens are indeed "contemporary classics" (mined within the last few decades), others, such as the wire silvers from Kongsberg, Norway (plates 68 and 69), were originally extracted from the earth more than 200 years ago and have survived the centuries to be enjoyed by present and future generations.

The following selection of photographs highlights 80 of the more than 500 mineral specimens currently on public view. Limiting the selection to 80 "mineral masterpieces" was quite difficult, because so many are not only extraordinarily beautiful and significant but also exceptionally photogenic. Those involved in the selection process frequently arrived at different preferences, testifying to the great wealth of specimen material that the collection has to offer. Visitors to the museum are invited to choose their own favorites, bearing in mind that only half the collection is currently on exhibit, and an active acquisitions program is adding regularly to this remarkable and unique archive of natural beauty.

For those who are unable to visit the Houston Museum of Natural Science we hope that the work you now hold in your hand will give you a glimpse of what this incredible collection has to offer. As such, the board of trustees and staff would like to thank several underwriters who made the publication of this book possible. Our gratitude for bringing *Masterpieces of the Mineral World: Treasures from the Houston Museum of Natural Science* to fruition is extended to: Mr. & Mrs. E. A. Bartsch, Jr.; The Cockrell Foundation; The Fruehauf Foundation; Mr. & Mrs. Henry R. Hamman; George & Catherine Lindahl; Mr. & Mrs. Herbert P. Obodda; and Martin Zinn III.

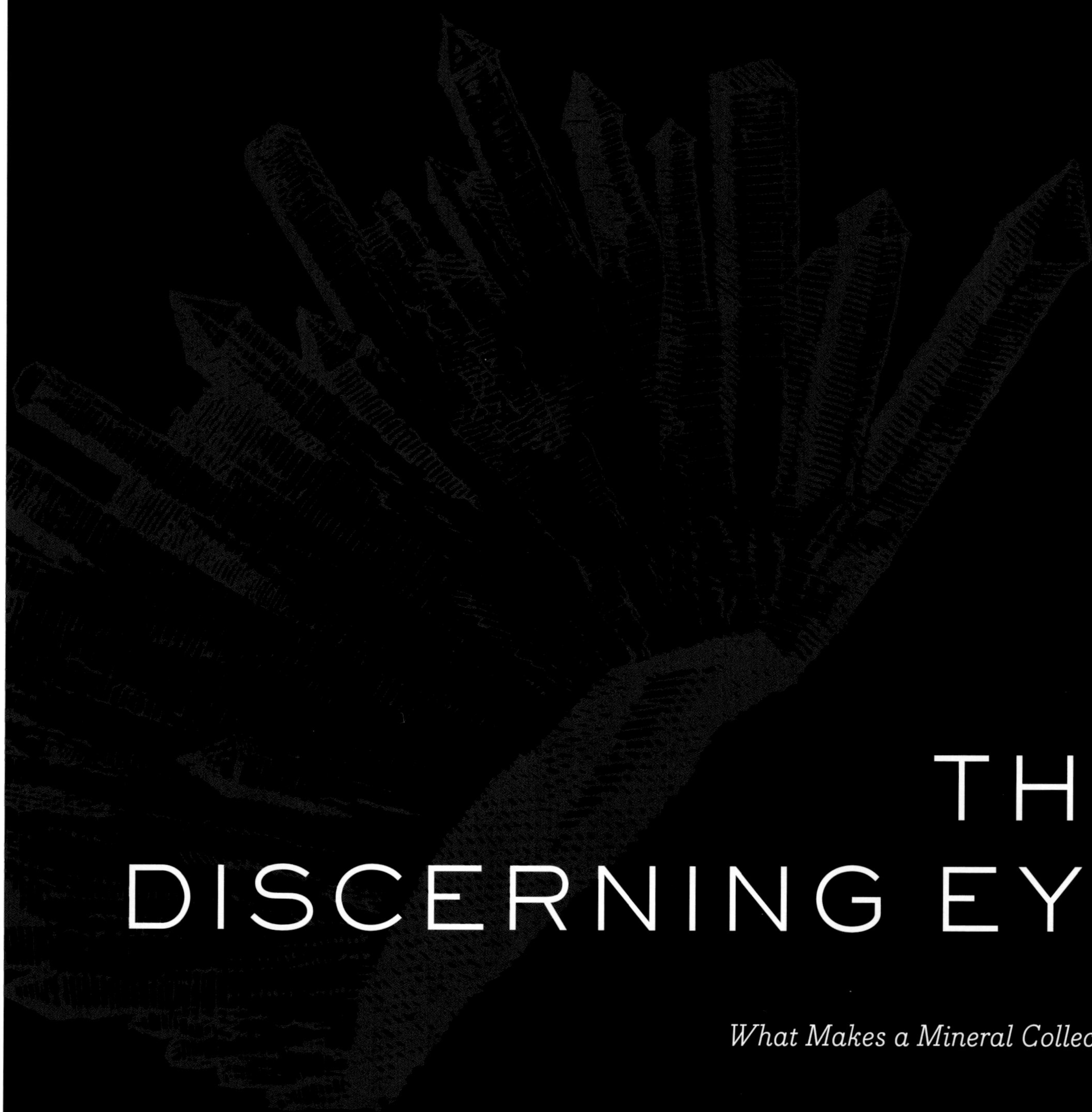

THE DISCERNING EYE

What Makes a Mineral Collectible?

Traditionally, a *masterpiece* is a creation that represents the finest work of a particular artist, working in a particular medium. The concept is equally applicable to the creations of nature that comprise the mineral world. Countless millions of crystals of a given mineral may form at various places in the earth, but at one or two special locations (for reasons that fascinate the mineralogist) the crystals will have had the opportunity to grow bigger and better in every way, than at any other site. The most perfect and aesthetically pleasing of these natural productions are the masterpieces of the mineral world.

The aesthetic appreciation of minerals requires no scientific knowledge at all; it is based purely on visual appeal. An appreciation of the technical aspects of a mineral specimen necessarily involves scientific and historical knowledge. An aesthete or artist sees minerals as the artworks of nature, whereas a mineralogist, trained in the concepts of crystallography, chemistry, and geology, sees in minerals a complex system of atomic relationships, which is a direct consequence or expression of the laws of physics and chemistry, in concert with rare combinations of geological and historical conditions. The finest quality mineral specimens exhibit the best presentation of physical, aesthetic, and scientific characteristics, so that the species may easily be distinguished from others as the most illustrative and illuminating examples of their kind.

This section will strive to distinguish those scientific advantages and those eternal aesthetic verities that allow taste to rise above temporary fashion. To do so, one must first identify the aesthetic qualities and scientific-historical criteria that are typically taken into consideration when selecting a truly great mineral specimen.

CRYSTAL F

Of all the criteria that might be app
complete or nearly complete cryst
manifestation of the internal arrang
composition, defines a mineral's id
of crystals, the rhythmic repetition c
well as the eye.

Ideally the crystal shapes sho
and minimal distortion, however th
(rounded, hoppered, covered with
fascinating and attractive specimer
present, the specimen has added s

COLOR

Perhaps the two most significant characteristics that some of the most beautiful minerals possess are purity of color and transparency. People derive pleasure from gazing into a colorful, transparent solid – it is a fundamental peculiarity of the human mind, producing an almost tranquilizing effect. Many mineral species are incapable of exhibiting these characteristics, the opaque metallic minerals for instance, but for any species that is capable of exhibiting bright colors and transparent crystals, it must do so in order to be considered an exceptionally fine specimen.

The crystals, of course, may well be on matrix (see section on Other Properties), and may also be accompanied by other mineral species. Consequently it is not just the color of the crystals that must be evaluated, but also the complementary matrix colors and the presence of the other minerals, as well as how well all of these colors contrast and harmonize with each other to produce an overall aesthetic effect.

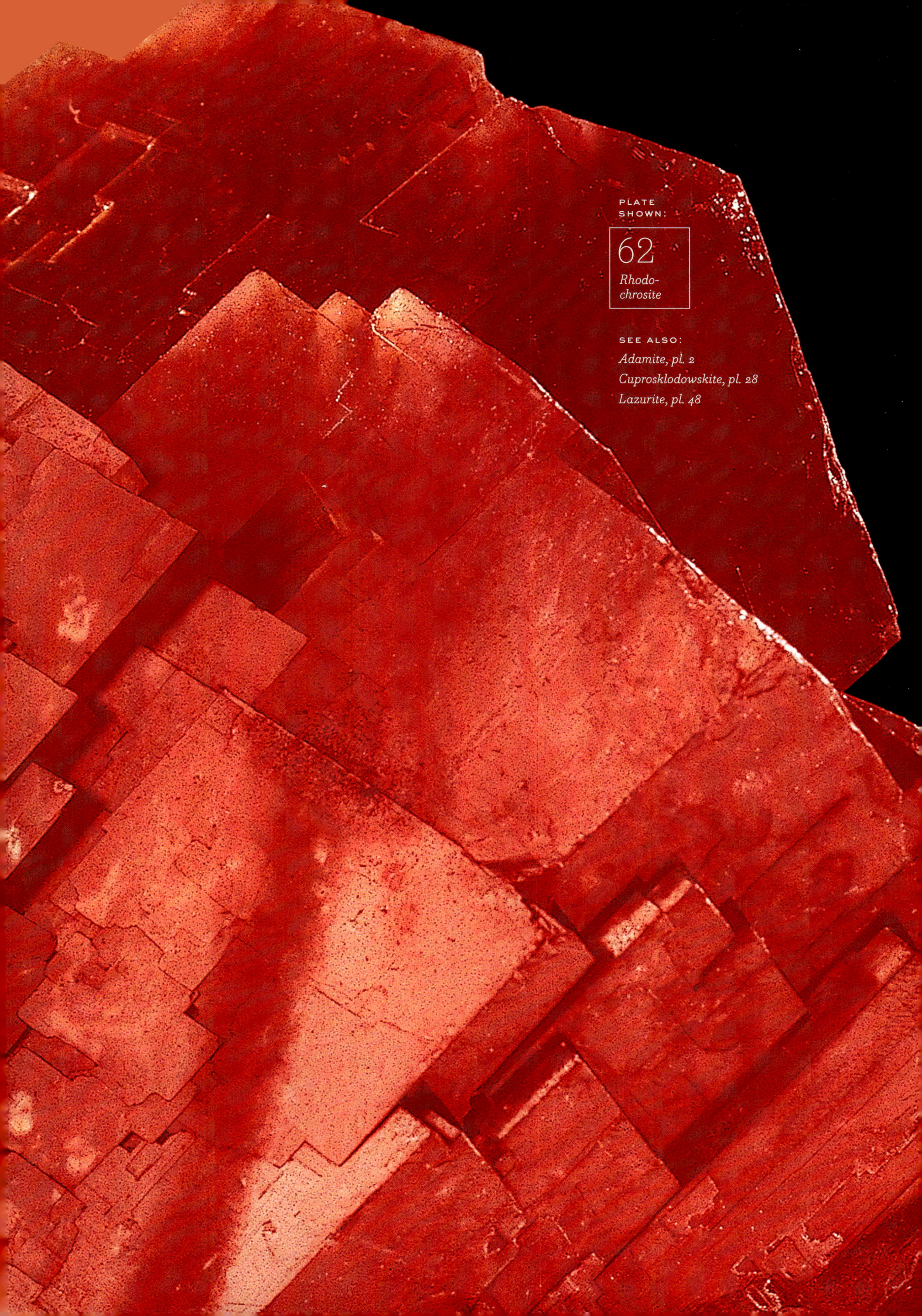

PLATE SHOWN:

62
Rhodochrosite

SEE ALSO:

Adamite, pl. 2
Cuprosklodowskite, pl. 28
Lazurite, pl. 48

TRANSPARENCY

Transparency gives evidence of internal perfection, as if the crystal interior had been swept clean of litter and dirt. It also requires a lack of internal damage (cracks), and a lack of growth flaws and inclusions. The most perfectly transparent crystals are referred to as *water-clear*, so perhaps the basic human need for pure drinking water is an underlying factor in our minds. In any case, the fundamental criteria for gemstones are color and clarity, so it should be no surprise that these are also important in the aesthetics of uncut, natural mineral crystals.

PLATE SHOWN:

10

Beryl

SEE ALSO:

Beryl, pl. 12
Cerussite, pl. 19
Spodumene, pl. 74

LUSTER

In addition to the importance of shape, the quality or nature of the surface of the crystal is a critical characteristic. "Luster" is a general term describing the overall reflective effect resulting from the fine-scale perfection of a crystalline surface, as well as from the refractive and absorptive properties characteristic of a particular species. Crystal surfaces, if perfect, appear to be absolutely flat all the way down to the microscopic level. Imperfections, or deviations from absolute flatness and "perfection," can occur due to a variety of causes including crystal growth defects, the presence of surface contaminants, or the dissolution of a previously formed crystal resulting from exposure to corrosive liquids or gases. Such irregularities reduce the amount of light that is reflected in a single direction from a face, reducing the brilliance of the reflection.

It should be remembered that in the case of transparent crystals, light can be reflected from the *interior* side of each face as well as from the exterior side, thus increasing the total amount of light being reflected back to the viewer. This effect can deepen the apparent color of a crystal because the effect is the same as if the crystal were thicker. Apparent luster and color may also be affected by changing the direction from which the crystal is being viewed.

PLATE SHOWN:

44

Hematite

SEE ALSO:

Proustite, pl. 57
Topaz, pl. 77
Zoisite, pl. 80

actual size

PLATE SHOWN:

55

Mimetite

SEE ALSO:

Azurite, pl. 3
Elbaite, pl. 31
Quartz, pl. 60

SIZE

Size is a surprisingly difficult aspect to evaluate in terms of fundamental appeal. All other things being equal, bigger is probably better. Unfortunately, all other things never *are* equal. Increasing size comes with an increasing tendency toward defects of all kinds. Micromounters (people who collect mineral specimens that are so small that they must be viewed under magnification in order to be fully appreciated) know well that the very best and most beautiful crystals of a great many species are under one or two millimeters in size. When size is balanced against the associated disadvantages relative to other criteria, people fail to agree on an optimum. Bigger may be better or more desirable, or certainly more rare, but beauty, in our opinion, exists independently of size.

In practice, the well-rounded connoisseur is able to appreciate and evaluate specimens of any dimension, applying all of the other criteria *in the context* of size.

PLATE SHOWN:

11

Beryl

SEE ALSO:

Beryl, pl. 10
Cerussite, pl. 19
Gold, pl. 42

PERFECTION

Damage is, of course, the antithesis of perfection, so much so that freedom from damage becomes a criterion of its own. Because truly perfect specimens are extremely rare, perfection is often the hallmark of a masterpiece. In rare instances, however, certain imperfections may offer relative improvement to other criteria such as internal reflectivity and color intensity. Similarly, with regard to luster, a crystal with alternating lustrous and frosty faces may be extremely aesthetic. In such cases the overall effect must be considered when making aesthetic judgments, and the fine sensibilities of the connoisseur are once again essential.

Repaired specimens are a special case. Where the repair is obvious, or where pieces are clearly missing, desirability must inevitably suffer. But where expert repairs are essentially invisible, many connoisseurs experience little diminution in their appreciation for the specimen. While unrepaired specimens are more highly desirable simply because they are more "perfect," an invisible repair does not affect aesthetics (which deals only with the visible), nor does it alter any of the existing scientific elements relating to the unseen aspects, parameters, and concepts of mineralogy.

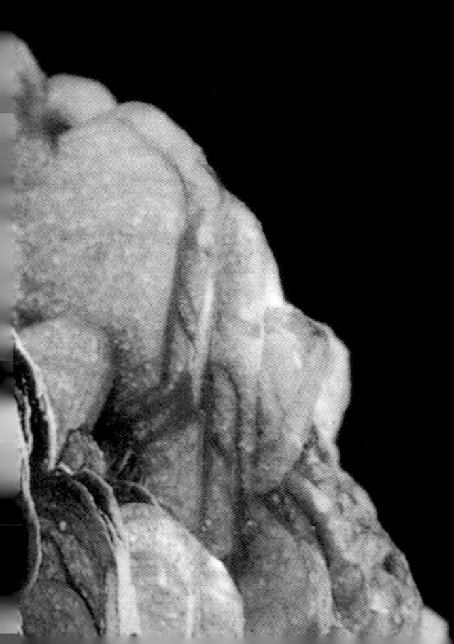

PLATE SHOWN:

18

Calcite

SEE ALSO:

Cerussite, pl. 20
Elbaite, pl. 33
Mesolite, pl. 52

CRYSTAL ORIENTATION *and* GROUPING

The orientation and grouping of crystals on their original matrix is critical to the overall aesthetic and scientific appreciation of a mineral specimen. Crystals standing in high relief (more or less perpendicular to their matrix), and separated or isolated somewhat from adjacent crystals, reflect light more aesthetically, show their shape better, and are more recognizable than those that are lying flat on the matrix or are thickly intergrown.

AESTHETICS

A broken or stream-rounded lump of mineral can possess color and transparency, but has very little coherent external design. To impress us aesthetically, the overall shape of crystals, crystal groups, and matrix, individually and in concert, must take on a pleasing *sculptural* quality. What makes one shape more aesthetically pleasing than another may have something to do with our subconscious appreciation of mathematics and geometry. A part of our mind is constantly engaged in making visual proportional estimates and comparisons on several levels. When numbers or ratios are found to repeat, resonate, or harmonize in novel ways, a sort of pleasure alarm goes off in our minds. This may, in part, be a parallel to the scientist's appreciation for a universe that is logically consistent and harmoniously interrelated. But to some extent it may also be inborn in the human mind, as evidenced by extensive cross-cultural studies showing general aesthetic preference for rectangles having a ratio of side lengths equaling 1:1.618 – the so-called *golden section.* Aesthetic ratios might involve the size of the crystals relative to the size of the matrix, or the length of a crystal relative to its width, or the way in which the crystals are arranged on matrix.

Aesthetically pleasing composition often requires some kind of focal point or center of interest in a specimen (perhaps, for example, the largest crystal in a cluster). The same is often true of any piece of artwork; the eye follows a path in examining it all, arriving at the focal point. The ultimate in sculptural aesthetics is found in those few mineral specimens that remain aesthetically beautiful regardless of the viewing angle. This property is rare even among many of the world's finest mineral specimens.

PLATE
SHOWN:

41

Gold

SEE ALSO:
Elbaite, pl. 31
Mesolite, pl. 53
Silver, pl. 69

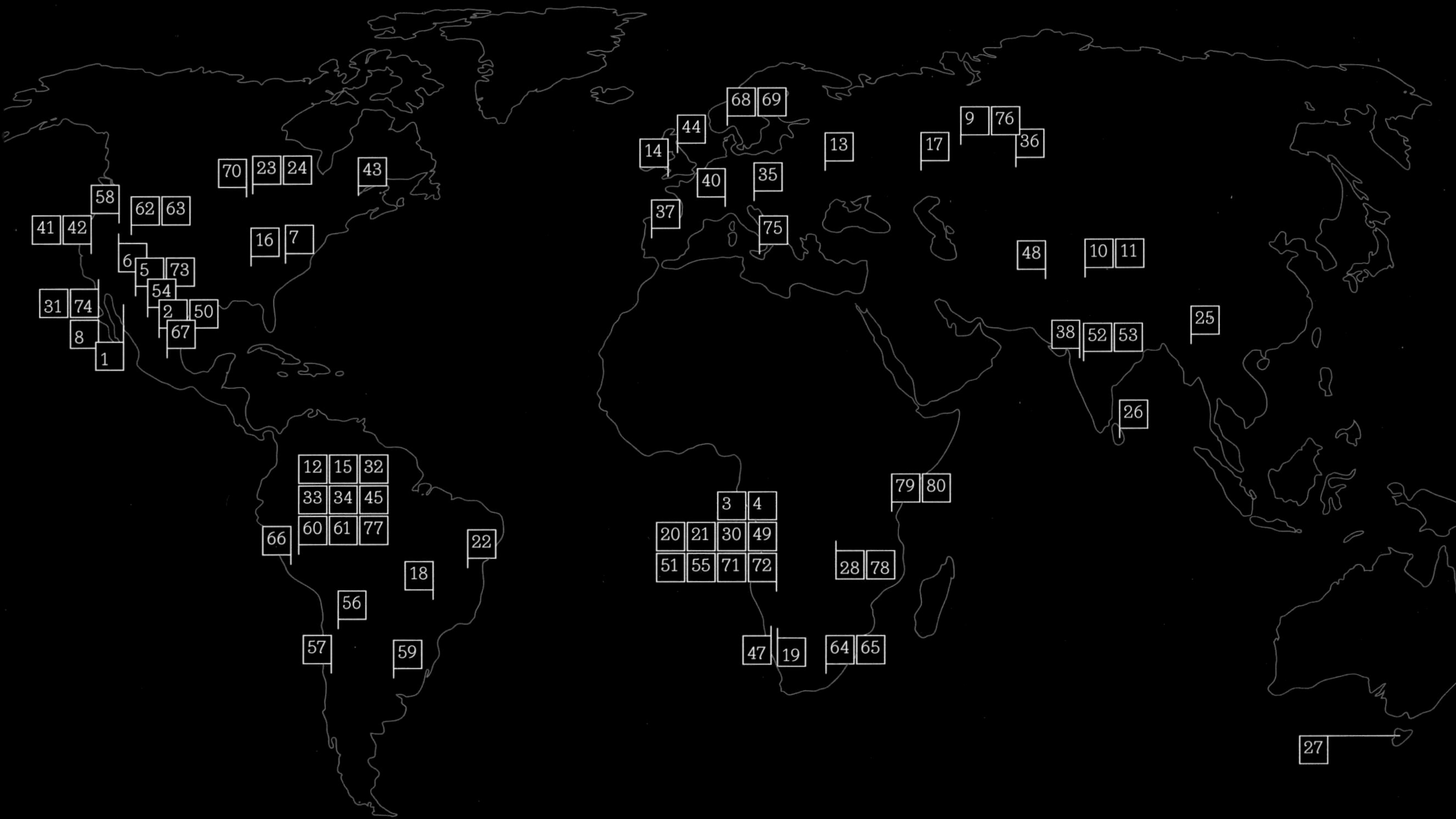

PROVENANCE

A specimen's provenance – its place of discovery and the history of its ownership since the time it was collected – is always of interest. Generally speaking, the more a person knows about the history of mineralogy and the history of mineral collecting, the more he will appreciate provenance. The provenance data accompanying a specimen can be correlated with its place in human history to yield insights and intellectual satisfaction. To know that a specimen was, for example, once a part of the collection of René Just Haüy (the founder of modern mathematical crystallography), or that it was collected during the Lewis and Clark expedition, can be just as enjoyable as any other aspect of mineral collecting.

OTHER PROPERTIES

Associated Species

Some minerals occur in association with other species that may or may not be as large and attractive, but the mere presence of associated species is an advantage because it shows a chemical and environmental relationship. In some cases those species may be trapped within large, transparent crystals of another species and can be seen floating inside. They may have formed on the surface of the crystal as it was growing, and now serve to delineate the earlier, smaller shape of the crystal. Such internal images (called "phantoms") or merely scatterings of small crystals (called "inclusions") can add special interest to any specimen. In some cases the mineralogist can work out detailed chemical explanations for why two minerals occur together, or one inside the other, lending scientific value to the specimens.

Durability

Durability is a practical consideration rather than a scientific or aesthetic one, but it is nevertheless critically important. Some minerals will decompose upon exposure to strong light or to fluctuations in heat and humidity, others may consist of a friable matrix that readily breaks apart, and some mineral specimens are so excessively fragile that they cannot be handled at all without being damaged. Ephemeral specimens such as these, though perhaps just as beautiful, will not be as highly coveted by museums and connoisseurs as those minerals that are capable of surviving for centuries when properly cared for.

Preparation

Some mineral specimens are perfect the moment they are extracted from the earth, others may require some level of "specimen preparation" after they have been mined or collected. Specimen preparation usually involves cleaning (to remove unwanted debris, stains, and other mineral coatings) and trimming (to adjust the crystal-to-matrix size ratio in an effort to refine the sculptural composition of the overall specimen). During the preparation process, some mineral specimens may also need to be repaired (see section on "Perfection"). Any *restoration* of a mineral specimen that involves the addition of foreign materials, dyes, fillers, etc., or any kind of modification of a specimen that employs lapidary techniques, renders such specimens less desirable than comparable specimens that have not been restored using such treatments.

Authenticity

In order to appreciate provenance the connoisseur must also be shrewd in his ability to critically evaluate a specimen's purported history. Simply taking someone else's word for it requires no discriminatory powers. Has the specimen passed exclusively through the hands of reliable and unbiased people? Can it be correlated with published descriptions or illustrations? Does the label appear authentic, and does it match catalog numbers that may be painted on the specimen? Can the handwriting on the label be recognized? Is the original catalog still extant and can an entry therein be matched with the specimen? Are letters or other supporting documents available? Many questions must be answered before important provenance is accepted. And many specimens have irretrievably lost their provenance. Although scientific aspects may often be rediscovered through new analyses, provenance, once lost, is likely never to be regained. Consequently, the connoisseur does all he can to verify and preserve labels and other provenance documentation.

Human nature being what it is, there is always the danger that some specimens will have been artificially modified so as to better meet the standards of the connoisseur. Luster, color, clarity and shape can all be altered to improve aesthetic appeal; even scientific aspects can be falsified (e.g., addition of foreign matrix or falsification of locality data). A critical, suspicious, and discerning eye is necessary for the connoisseur to avoid being fooled. It is of little use to have highly refined scientific and aesthetic criteria if those criteria can be easily circumvented by unscrupulous people who would deceive the collector. Protection requires knowledge and experience, so that very subtle deviations from the natural state can be recognized.

Matrix

Regardless of how beautiful a crystal may be, the connoisseur wants to see it still attached to the original piece of rock on which it grew (the *matrix*). The matrix, though perhaps not particularly aesthetic by itself, preserves geological knowledge by showing something of the environment in which the crystal developed, and therefore makes for a more complete and intellectually satisfying specimen. Preservation of matrix is not always possible, but is strongly preferred.

Rarity

Rarity, though not necessarily an aesthetic factor, is a necessary criterion to be considered when determining whether one mineral specimen is superior to another. Rarity is a circumstance, an outside factor, and perhaps even a temporary condition when viewed in light of the fact that new discoveries can render once rare specimens relatively common. The rarity of an individual specimen can change drastically without the specimen itself being physically altered in any way.

Of course, certain chemical *elements* occur only rarely in the earth's crust, and to some extent the mineral species that contain those elements may also be considered rare, but there are also many extremely rare species composed of very common elements. Cosmic elemental abundance is an unreliable standard for appreciating the rarity of minerals.

Rarity can be evaluated from many different viewpoints, including (1) rarity of the species, (2) rarity of the species from a certain locality, (3) rarity of the species having a certain crystal habit or color, (4) rarity of specimens having a certain provenance or history, (5) rarity of a species in well-crystallized specimens, and so on. Perhaps the most common approach to rarity is based on the *degree* to which specimens satisfy one of the various scientific and aesthetic criteria. Naturally the biggest, most richly colored, most lustrous, most transparent, or most sculpturally pleasing specimens are virtually by definition rare, although proximity to the next specimens in rank makes a significant difference in how that rarity is weighed.

The evaluation of rarity is, in part, simply a statistical exercise: "How many specimens have I seen that are as good, or better, than this one?" Of course, this is somewhat subjective since the greater the viewer's personal databank of experience, the more significant the answer. Somewhat confusingly perhaps, rarity is also linked with size and perfection, as well as with aesthetic, historical, and scientific criteria in ways that can often make it difficult or impossible to isolate. In terms of *desirability*, a specimen that is as ugly as dirt and carries no scientific interest benefits hardly at all from also being rare (or, for that matter, from being large or having an interesting provenance), but a very beautiful specimen benefits tremendously. Rarity, provenance, and size tie the various scientific and aesthetic criteria together to form an interactive complex. They function as desirability enhancers, and their effect is more or less proportional to a specimen's scientific or aesthetic desirability.

PLATES

1 ACANTHITE

LAS CHISPAS MINE, ARIZPE, MUNICIPIO DE ARIZPE, SONORA, MEXICO Acanthite (silver sulfide) is one of the most important ore minerals of silver. In rare cases it is found in skeletal crystal growths such as the magnificent 9.3-cm cluster shown here. This example is unusually large and perfect, with interesting parallel crystallizations and an overall attractive shape. Bits of matrix and associated minerals adhere at the bottom.

2 ADAMITE

SAN JUDAS CHIMNEY, LEVEL 6, OJUELA MINE, MAPIMI, DURANGO, MEXICO Adamite comes in many colors, but the finest specimens have always come from the Ojuela mine, and by far the most sought-after color is the royal purple found in the San Judas chimney in 1981. The mineral world was galvanized by the extraordinary specimens recovered from that find. Perhaps the best of the lot is this 5.5-cm cluster on limonite matrix – making it the best known example of the species. The complexity, sharpness, perfection, color, arrangement, and luster of the semi-transparent crystals are all superb.

3 AZURITE

TSUMEB MINE, TSUMEB, NAMIBIA The Tsumeb mine has produced the world's finest azurite crystals, of which this large 11-cm crystal group on green smithsonite is one of the best examples. The highly lustrous, elongated crystals with perfect terminations, on a contrasting base, admirably fulfill the requirements of connoisseurship.

4 AZURITE

TSUMEB MINE, TSUMEB, NAMIBIA These two exquisite and well-formed 5-cm crystals on matrix are perfect – a prime example of azurite from the Tsumeb mine. The highly lustrous, blocky, sharp, undamaged crystals resting separately on a contrasting matrix represent one of the most aesthetic and sought-after habits of Tsumeb azurite.

5 AZURITE

HOLBROOK MINE, BISBEE, COCHISE COUNTY, ARIZONA Botryoidal malachite is well-known from the famous Bisbee copper mines but botryoidal azurite, especially such a fine specimen as this 7.6-cm example from the Holbrook mine, is relatively rare. The aesthetic shape is particularly appealing.

6 BARITE

MEICKLE MINE, ELKO COUNTY, NEVADA The Meickle mine has produced some of the world's finest barite specimens, in beautiful clusters of gemmy, bright yellow, highly lustrous crystals of large size. The dominant 4.2-cm crystal shown here, collected on the 328-meter level sometime between 1996 and 1998, is unusual because it is isolated and on matrix, rather than part of a detached cluster.

7 BERYL

variety EMERALD

NEAR HIDDENITE, ALEXANDER COUNTY, NORTH CAROLINA Unearthed in December 2003, this rich green emerald crystal from North Carolina weighs approximately 1,869 carats making it the largest emerald crystal ever recovered in North America. While the size of the crystal is notable, even more remarkable is the fact that it is, by several orders of magnitude, the finest quality emerald crystal ever recovered in North America, and, by virtue of its overall aesthetic beauty, considered by many connoisseurs as the finest emerald specimen ever found anywhere in the world.

8 BERYL

variety MORGANITE

WHITE QUEEN MINE, SAN DIEGO COUNTY, CALIFORNIA The White Queen mine is famous for beautiful crystals of pink beryl (the gem variety "morganite," named after financier J. P. Morgan). The pictured 7.9-cm crystal, found in 1964, may well be the best ever found there. It has good color, no damage, a glassy luster on the big hexagonal "c" face, and interesting crystal form, and it rests on a base of white albite crystals. Peter Bancroft, in his *World's Finest Minerals and Crystals* (1973), considers this specimen to be the world's finest example of pink beryl (although fans of examples from Madagascar and Brazil might put up an argument).

9 BERYL

variety HELIODOR

MURSINKA, ALABASHKA-MURSINKA PEGMATITE FIELD, URAL MOUNTAINS, YEKATERINBERG OBLAST, RUSSIA

Mursinka heliodor (yellow beryl), of which this superb 8-cm, beautifully terminated crystal is a prime example, is one of the great classics among the gem minerals. The Ural Mountains deposits have produced very little during the last century, so most surviving uncut crystals are very old and quite rare.

10 BERYL *variety* AQUAMARINE

DUSSO, NORTHERN AREAS, PAKISTAN Aquamarine crystals from northern Pakistan have been known for nearly a century, but superbly gemmy, lustrous, beautifully formed crystals on white feldspar matrix such as the one pictured have always been rare. Especially unusual is the large size of the crystal, 13 cm, versus its remarkable clarity, as larger crystals tend to have more internal flaws and milkiness.

11 BERYL

DUSSO, NORTHERN AREAS, PAKISTAN Over the course of the last several decades Pakistan has produced what are arguably the world's finest crystals of aquamarine. What makes these pieces particularly remarkable is the fact that the crystals often remain attached to the original matrix upon which they formed. This piece consists of an near-flawless 8 cm aquamarine crystal surmounting a matrix of silvery muscovite crytals. The perfection, color, and clarity of the crystal, along with the crystallized matrix, mark this aquamarine among the finest ever recovered from Pakistan.

12 BERYL

variety GOSHENITE *on* MICROCLINE

MINAS GERAIS, BRAZIL This superb 13.6-cm crystal of colorless beryl shows unusually sharp edges and smooth faces. The association with feldspar is an advantage as well, because in many Brazilian pegmatites the feldspar has weathered away to clays leaving nothing to serve as matrix for the gem crystals. This specimen is from an unknown pegmatite in the Brazilian interior.

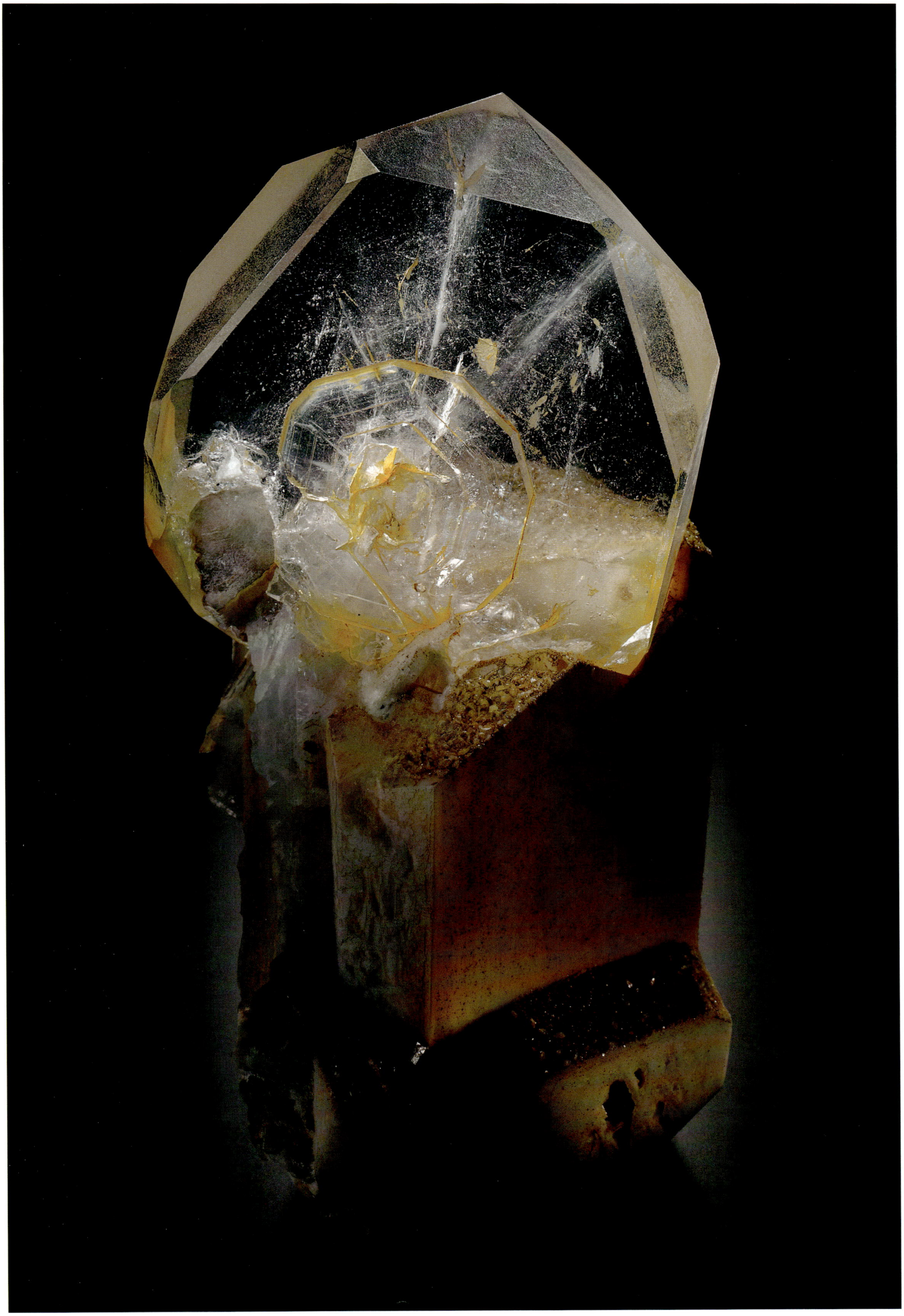

13 BERYL

variety HELIODOR

VOLODARSK VOLYNSK PEGMATITES, ZHITOMIRSKAYA OBLAST, UKRAINE Gem beryls from the Ukrainian pegmatites were not seen in the West until some crystals began to be smuggled out in the late 1970s. Following the fall of Soviet Communism around 1989, more were exported. The large, 14-cm crystal pictured here, from a major pocket found in 1992, is a superb example, showing the typically etched but highly lustrous surface.

14 BOURNONITE

HERODSFOOT MINE, LISKEARD, CORNWALL, ENGLAND Cogwheel-like groupings of crystals of bournonite from the Herodsfoot mine have been treasured as the finest examples of the species since they were found there by miners from 1858 to 1868. The pictured example, measuring 4.7 cm, has sharp, lustrous, undamaged crystals (with an attractive blue tarnish in places) perched on a brilliant white contrasting matrix of sparkling quartz crystals.

15 BRAZILIANITE *with* MUSCOVITE

CÓRREGO FRIO NEAR LINOPOLIS, MINAS GERAIS, BRAZIL The world's finest specimens of the mineral brazilianite came from the type locality, the first place where it was ever found, despite numerous later discoveries of the mineral in other deposits nearby. The beautiful 5.5-cm specimen shown here is a prime example, exhibiting perfect form and color with good transparency, no damage, and some attached crystals of muscovite. It was collected in 1945–1946 by the American dealer/collector Edward Swoboda, who recovered most of the finest specimens from the occurrence.

16 CALCITE *on* BARITE

ELMWOOD MINE NEAR CARTHAGE, SMITH COUNTY, TENNESSEE Golden yellow to reddish yellow calcite crystals on barite were the largest and most beautiful products of the Elmwood mine. Retrieving large and delicate calcite crystals was always a very difficult task in a working mine, and the vast majority suffered at least a small amount of damage, especially to the sharp tips, which easily cleave off. The 10-cm crystal pictured here is rare for its high degree of perfection, and it shows excellent luster, interesting crystal form (it is twinned), and nice color.

17 CALCITE

SARBAYSKAYA QUARRY, RUDNIY, KUSTANAY OBLAST, KAZAKHSTAN The Sarbayskaya quarry, adjacent to the Sarbay iron mine in Kazakhstan, has been a source of large and beautiful golden orange calcite crystals since the early 1990s. The blocky orange-yellow crystals are abundant, and many specimens have been collected that are completely free of damage. The 8.6-cm example pictured here, however, is particularly large and fine. The interesting crystal surfaces, silken luster, rich color, and good transparency make it a superb specimen worthy of any collection.

18 CALCITE

IRAI MINE, IRAI VILLAGE/STATE, RIO GRANDE DO SUL, BRAZIL Calcite is a common mineral found all over the world, but this specimen of milky white crystals in a 14.4-cm cluster on matrix is a masterpiece. The soft luster of the crystals and the aesthetic way in which they are grouped on matrix make this a stunningly attractive and sculptural piece regardless of the abundance of the species.

19 CERUSSITE

KOMBAT MINE, OTAVI, NAMIBIA This exquisitely brilliant, twinned 6-cm crystal of cerussite is the finest known example of its species – nature's version of "lead crystal." The highly lustrous, perfectly transparent pair of crystals is almost 80% lead, the remaining 20% consisting of carbon and oxygen. The play of colors seemingly trapped within the crystal results from cerussite's high refractive index and its exceptional clarity. This specimen was recovered from a mine located just 49 km south of the famous Tsumeb mine in north-central Namibia.

20 CERUSSITE

TSUMEB MINE, TSUMEB, NAMIBIA Cerussite is among the most famous minerals found at the Tsumeb mine. Countless thousands of specimens have emerged, but the reticulated, twinned groups are especially interesting. Formed by the intersection of three sets of crystals locked together crystallographically at 60° angles to each other, the result looks like a giant snowflake. The 26-cm reticulated cluster shown here is one of the largest and best ever found.

21 CERUSSITE

with DOLOMITE

TSUMEB MINE, TSUMEB, NAMIBIA This superb 9-cm cerussite consists of crystals of the elongated habit joined together by twinning to form a remarkably aesthetic and sculptural specimen. The white dolomite crystals sprinkled along the crystal arms also add interest, and the specimen even retains some matrix. It was sold by the dealers Miriam and Julius Zweibel, who were among the first to bring large quantities of Tsumeb minerals to the American market in the early 1970s.

22 CHRYSOBERYL

(SIXLING TWIN)

ITAGUAÇU, ESPIRITO SANTO, BRAZIL The finest and most desirable chrysoberyl specimens are twinned in a six-armed pinwheel shape by the interpenetration of three individuals at 60° angles to each other. Large and beautiful specimens are known principally from a remote occurrence in Espirito Santo State in Brazil, the 6.5-cm example pictured here being a prime example, with good form, good size, deep color, partial transparency, no damage, and excellent luster. It was collected in 1940.

23 COPPER

KEWEENAW PENINSULA, HOUGHTON COUNTY, MICHIGAN Large single crystals of copper are extremely rare, despite the vast tonnages of native copper that were produced in Michigan's "Copper Country" since mining began there in the 1840s. At 5.7 cm, the example shown here may be the largest, high-quality, single crystal known. It is also undamaged, shows interesting surface growth features, and has an aesthetic group of smaller crystals attached.

24 COPPER

KEWEENAW PENINSULA, HOUGHTON COUNTY, MICHIGAN Large clusters of free-growing, blocky copper crystals like this 18.4-cm specimen are extremely rare. The example pictured here appears to be from the same mine as a superb 13-cm specimen known to have come from the collection of B. S. Butler who (with W. S. Burbank) wrote the classic work on the copper deposits of Michigan in 1929. As with many such early specimens, the specific mine name was lost or not recorded. Nevertheless, this is one of the finest copper specimens known from Michigan.

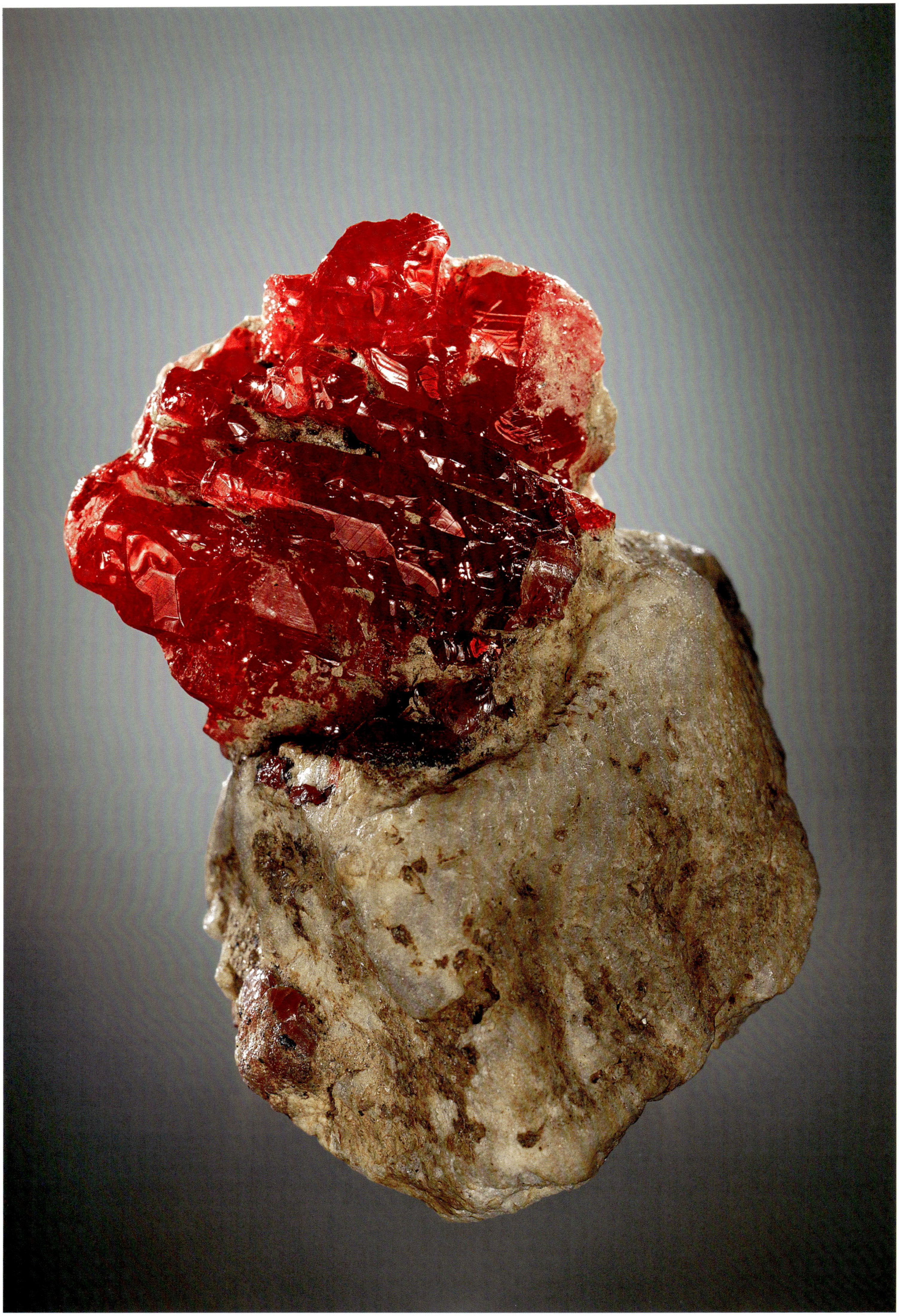

25 CORUNDUM *variety* RUBY

MOGOK STONE TRACT, KATHA DISTRICT, MYANMAR (FORMERLY BURMA) Mogok has produced the world's finest rubies for over 800 years. Gem-quality Burmese rubies are usually rather small; the 5-cm crystal on matrix shown here is exceptionally large, with bright luster and excellent "pigeon's blood" red color. Gem crystals such as this one that still have matrix attached are very rare.

26 CORUNDUM *variety* SAPPHIRE

RATNAPURA, SRI LANKA Sapphire crystals like the beautiful 11-cm specimen pictured here are typically waterworn; only when they are collected very near to their original hard-rock source area are they so sharp-edged and unworn. The size, attractive yellow-to-blue coloring, excellent luster, sharpness, and transparency make this crystal a truly rare and superb example of Sri Lankan sapphire.

27 CROCOITE

ADELAIDE MINE, DUNDAS, TASMANIA, AUSTRALIA Despite its relatively small size (4.5 cm), this pair of crocoite crystals found in the 1920s are considered by some to be the finest example of the species. The crystals individually are comparatively quite large and they are both perfectly terminated. The small amount of matrix makes the specimen even more interesting.

28 CUPROSKLODOWSKITE

MUSONOI MINE NEAR KOLWEZI, KATANGA, DEMOCRATIC REPUBLIC OF THE CONGO Pockets in uranium-bearing ore lined with gorgeous, yellow-green, needle-like crystals of cuprosklodowskite have been very popular with mineral collectors since the 1950s, when great specimens like the 7-cm example pictured here were discovered at the Musonoi mine. Although the mineral was first found at the neighboring Kalongwe mine, it reaches its highest development at Musonoi.

29 DIAMOND

MIR PIPE, YAKUTIA, SIBERIA, RUSSIA The 2-cm gem-quality diamond crystal pictured here is surprising for its large size and for the presence of original matrix. The Russian government has traditionally reserved such large crystals for its national treasury. The Mir pipe in Siberia has been unusually prolific in the preservation and sale of diamonds in matrix, but more commonly the crystals are in the 1 cm range.

30 DIOPTASE

TSUMEB MINE, TSUMEB, NAMIBIA Dioptase is probably the most famous of all minerals for which Tsumeb is known. Large numbers of specimens reached the market in the 1970s, but crystals much larger than 1 or 2 cm were always rare. Because dioptase crystals often form intergrown clusters of crystals, the large 4.1-cm crystal pictured here is remarkable because it is a single undamaged crystal perched on a crystallized matrix of white aragonite. The deep color, high luster, and attractive crystal shapes make Tsumeb dioptase one of the most sought after of all mineral species.

31 ELBAITE *on* QUARTZ

TOURMALINE QUEEN MINE, SAN DIEGO COUNTY, CALIFORNIA North America has produced some extraordinarily beautiful specimens of elbaite, a member of the tourmaline group, but the most admired are the bright red-pink crystals with blue caps found in 1972 at the Tourmaline Queen mine. The 24-cm example pictured here is the finest of the 33 major specimens recovered and is therefore the finest North American tourmaline. The lustrous, lusciously colored, undamaged pair of crystals at the top grow from an undamaged quartz crystal and are accompanied by smaller tourmaline crystals. It has been nicknamed "The Rabbit Ears."

32 ELBAITE

SALINAS MINE, ARAÇUAÍ-SALINAS DISTRICT, MINAS GERAIS, BRAZIL Single gem crystals of elbaite, a tourmaline group mineral, such as this are expensive but eagerly sought-after prizes in the mineral world. The gorgeous red color, high luster, well-formed termination, and adhering bits of associated minerals at the base of this superb 9.5-cm specimen, not to mention the high degree of transparency and therefore gem value, make it a first-rate connoisseur's piece. Gem crystals of this quality are usually faceted into cut stones, so it is unusual to see one preserved as a crystal specimen.

33 ELBAITE

on QUARTZ

PEDERNEIRA MINE, SAN JOSE DA SAFIRA, MINAS GERAIS, BRAZIL Perhaps the most recent great discovery of elbaite has taken place at the Pederneira mine near San Jose da Safira, Brazil. The superb 14.8-cm group of green and pink crystals on quartz shown here is a prime example, showing deep color, high transparency, and luster.

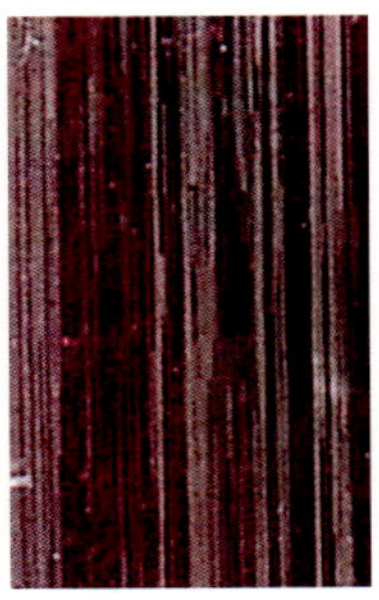

34 ELBAITE

JONAS MINE, ITATIAIA, CONSELHEIRO PENA, MINAS GERAIS, BRAZIL Cranberry-red elbaite from the Jonas mine represents one of the great collecting discoveries in the history of mineralogy. The specimens, like the 16-cm doubly terminated crystal shown here, have an exquisitely rich color, beautiful transparency, lustrous faces, attached matrix, and a very high degree of perfection. (In fact, the pocket was so clean when found that the crystals did not even need washing.) Although much larger crystals were recovered, most were cut for their gem value. Matrix specimens like the one shown are highly coveted today.

35 EPIDOTE *on* CALCITE

KNAPPENWAND, UNTERSULZBACHTAL, AUSTRIA The Knappenwand locality in Austria has produced the ultimate in fine epidote crystals since its discovery in 1865. The crystals are sharp, highly lustrous, and often come on matrix associated with a variety of other minerals. Though comparatively small, the 5.5-cm specimen pictured here shows all of the quality that has made specimens from this locality famous for 139 years. Fine examples continue to be very rare on the market.

36 FERRO-AXINITE

PUIVA MINE, SUBPOLAR URAL MOUNTAINS, TYUMEN OBLAST, CHANTY-MANSIJSK AUTONOMOUS REGION, RUSSIA

Ferro-axinite from the Puiva mine in the remote Subpolar Urals is by far the finest in the world. The best examples recovered, like the sharp, lustrous, gemmy, attractively colored, 5.9-cm crystal on matrix shown here, were immediately recognized as connoisseur-level pieces and were purchased by the major museums and private collectors.

37 FLUORAPATITE

PANASQUEIRA MINE, PANASQUEIRA, CASTELO BRANCO DISTRICT, BEIRA BAIXA PROVINCE, PORTUGAL

Purple fluorapatite is one of the most desirable minerals from the world-famous Panasqueira tin-tungsten mines. Specimens have appeared rarely on the market since the late 1960s or earlier. The 8.5-cm crystals shown here are exceptionally large, with great luster, color, form, and freedom from damage.

38 FLUORAPOPHYLLITE

MOMIN AKHADA NEAR RAHURI, MAHARASHTRA, INDIA The specimen pictured here is from a 2002 discovery. Green fluorapophyllite, one of India's most attractive and sought-after minerals, occurs in spherical clusters on a white stilbite matrix. This 12-cm example is among the largest and finest recovered. The spherical growth of such attractive crystals had been previously unknown in India and was greeted by much excitement in the mineral world.

39 FLUORITE

NIKOLAEVSKIY MINE, DALNEGORSK, PRIMORSKIY KRAY, RUSSIA Dalnegorsk is famous for producing large and superb fluorite crystals, some of which (like this remarkable 8-cm example) are so water-clear that they almost seem to disappear. From no other locality in the world do crystals occur as large and perfect, both internally and externally, as they do at Dalnegorsk.

40 FLUORITE

on SMOKY QUARTZ

NORTH FACE OF LES DROITES, MONT BLANC RANGE NEAR CHAMONIX, HAUTE-SAVOIE, FRANCE Pink fluorite from the French Alps is often associated with quartz crystals, but the amazing find pictured here, with dozens of sharp, richly colored crystals perched on an 18-cm quartz crystal, is rare. It is the best of three such pieces found by the French strahlers J. F. Charlet and R. Ghilini in an ice-filled cleft at an elevation of 3,600 meters in 1989.

41 GOLD

(NAMED "THE DRAGON")

COLORADO QUARTZ MINE, MARIPOSA COUNTY, CALIFORNIA The Colorado Quartz mine is one of California's most famous sources of beautifully crystallized gold. The lustrous, sheet-like clusters of sharp gold crystals grew around quartz crystals in open pockets and exhibit attractively sculptural shapes when liberated from their tight quarters. The large and superb 18.5-cm piece shown here has been nicknamed "The Dragon" because of its shape.

42 GOLD

MOTHER LODE COUNTRY, CALIFORNIA For sheer sculptural quality this 5-cm group of flattened gold crystals on quartz matrix from an unknown mine in California's Mother Lode country ranks among the most desirable gold specimens for the connoisseur. The rich yellow color, the mirror-bright luster, and the sharp crystal edges are all ideal for the species, and the lack of damage coupled with the presence of matrix combine to make it virtually a perfect specimen.

43 GROSSULAR

JEFFREY MINE, ASBESTOS, QUEBEC, CANADA Grossular, a garnet, is not an uncommon species, but perfectly transparent, highly lustrous, beautifully colored, dodecahedral crystals are extremely rare. The Jeffrey mine has produced the world's finest specimens, of which this 7.9-cm piece is a prime example. The interesting matrix, rich in associated species, makes this specimen particularly desirable.

44 HEMATITE

FLORENCE MINE, EGREMONT, WEST CUMBRIA, ENGLAND The aesthetically sculptural 7.8-cm specimen illustrated here is surely among the finest surviving examples of black hematite known as "kidney ore." The extraordinarily bright luster and deep black color (some botryoidal hematites from other mines are reddish instead of jet-black) distinguish this piece.

45 HYDROXYL-HERDERITE

XANDA MINE, VIRGEM DA LAPA DISTRICT, MINAS GERAIS, BRAZIL Hydroxyl-herderite from the Xanda mine is the finest in the world, not only because of its attractive purple color, but also because of the size and perfection of its crystals. The sharp and lustrous example shown here, a 7.3-cm crystal on white feldspar matrix, was found in 1976.

46 ILVAITE

FIRST SOVIETSKIY MINE, DALNEGORSK, PRIMORSKIY KRAY, RUSSIA Ilvaite from Dalnegorsk, Russia, is the world's finest, in lustrous black crystals on contrasting white quartz. The 6-cm crystal pictured here, from the Dalnegorsk occurrence producing the best specimens (the First Sovietskiy mine), is among the best, with interesting, complex faces, a high degree of perfection, and a formal orientation standing straight up on matrix. Dalnegorsk ilvaite was abundant on the market for a short time, just a year or two, and then more or less disappeared, indicating that the crystal-producing zone has probably been exhausted.

47 JEREMEJEVITE

MILE 72 NEAR SWAKOPMUND, NAMIBIA Blue jeremejevite is among the rarest of gem minerals and has been found at only two localities, both of them in Namibia. Crystals from the original Namibian find in 1973 remain the finest known examples of the species. The illustrated 5.5-cm specimen is the best of the matrix specimens recovered. The presence of multiple crystals makes it unusual.

48 LAZURITE

SAR-E-SANG MINE, KOKSCHA VALLEY, FIRGAMU, BADAKSHAN PROVINCE, AFGHANISTAN Large, well-formed crystals of deep blue lazurite partially embedded in white calcite have been coming out of Afghanistan since the 1980s, but most crystals are damaged when broken out of the enclosing rock. Specimens such as the 4.5-cm crystal illustrated here, with a large and perfect crystal, are highly unusual.

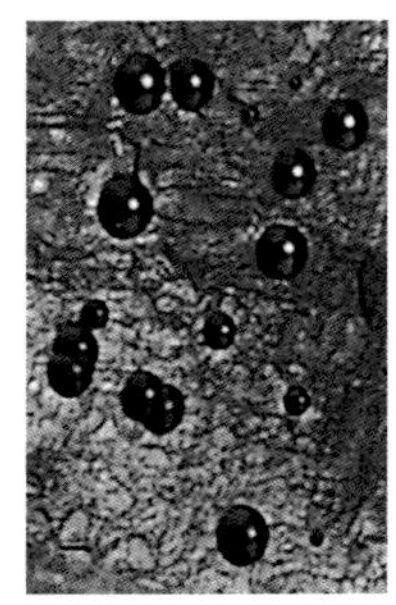

49 LEADHILLITE

TSUMEB MINE, TSUMEB, NAMIBIA The 20 world's finest leadhillite specimens were all found in the second oxide zone at Tsumeb in the 1960s. Flat hexagonal crystals up to 13 cm were found, including one crystal that is water-clear. Steep rhombohedral crystals like those on this 6.7-cm example shown are extremely rare – this is perhaps the only one of note. The tiny black globules are melanotekite.

50 LEGRANDITE

OJUELA MINE, NEAR MAPIMI, DURANGO, MEXICO This slightly divergent parallel cluster of legrandite crystals 10.7 cm long is from the famous 1977 pocket discovered by Felix Esquivel. It is from a group that collectively represents the finest examples of the species known. The huge size, rich yellow color, excellent terminations, and complete freedom from damage are especially appealing.

51 LUDLOCKITE

TSUMEB MINE, TSUMEB, NAMIBIA Ludlockite was originally found in a single boulder on the dump of the Tsumeb mine. It was described as a new species in 1972, based on that single occurrence. Subsequently a second, very limited occurrence was found deeper in the mine. Consequently very few specimens of the species are known. This beautiful, reddish-brown, 4-cm spray on zincian siderite crystals is among the finest.

52 MESOLITE *with* FLUORAPOPHYLLITE

PASHAN HILLS NEAR KHADAKVASLA, POONA, MAHARASHTRA, INDIA Mesolite in large crystals and large, spherical sprays like the one pictured have come from only one place in the world: the Pashan Hills quarries. The 23-cm example shown here, on fluorapophyllite matrix, is probably the world's finest example of the species. This specimen is extremely fragile and was brought out of India by mineral dealer Rock Currier, who packed the specimen in powdered soap and bought for it a first-class airplane ticket from Bombay to Los Angeles in order to ensure its safe arrival.

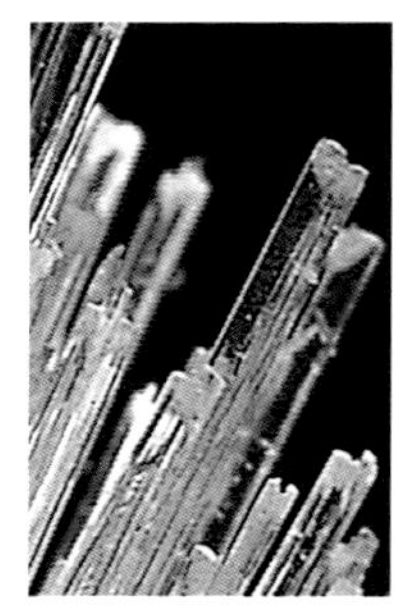

53 MESOLITE

with FLUORAPOPHYLLITE

PASHAN HILLS NEAR KHADAKVASLA, POONA, MAHARASHTRA, INDIA Attractive sprays of white mesolite on beautiful green fluorapophyllite, such as the 12-cm specimen shown here, come only from the Pashan Hills quarries. The Makki brothers, dealers in Indian mineral specimens, supervised the mining and preparation of the fragile specimens; without them, surely none would have been saved, and the world's greatest occurrence of mesolite would have been lost.

54 MIMETITE

CONGRESO-LEON MINE, SAN PEDRO CORRALITOS, CHIHUAHUA, MEXICO The 11.7-cm specimen of botryoidal mimetite shown here is among the best pieces collected from a single discovery made in 1968. Its complete freedom from damage, rich yellow color, sparkling silken luster, and aesthetic sculptural form qualify it as a connoisseur's specimen. Gemmy mimetite crystals (see plate 55) from Tsumeb, Namibia, are considered the finest examples of the species. Botryoidal specimens like this are so different in appearance, however, that they tend to be judged separately, and so the best examples of the San Pedro Corralitos find are treasured for their own unique beauty.

55 MIMETITE

TSUMEB MINE, TSUMEB, NAMIBIA Obtaining a specimen from the Tsumeb mimetite pocket of 1971 has been an obsession for many mineral collectors ever since the discovery. Only a small number of specimens, around 30, were recovered, but they show the most perfect quality known for the species. The crystals are water-clear and mirror-faced, with pleasant lemon-yellow color, simple crystal form, and are aesthetically clustered on matrix. The example pictured here measures only 2.9 cm but shows all of the qualities that make a world-class specimen.

56 PHOSPHOPHYLLITE

UNIFICADA MINE, CERRO RICO DE POTOSÍ, POTOSÍ, BOLIVIA Phosphophyllite crystals from Potosí, with their beautiful bluish green color, brilliant luster, and attractive transparency, are among the most highly desired treasures in the mineral world. They are rare today because most crystals were destroyed by mining before their identity was even understood. Any size crystal larger than 1 cm is highly valued, and this 6.8-cm twinned pair of gem crystals, the second largest known, should probably be considered priceless.

57 PROUSTITE

DOLORES TERCERA MINE, CHAÑARCILLO, CHILE Proustite is the rarest, most beautiful, most richly colored and most highly coveted of the "ruby silvers," silver minerals that are red and sometimes transparent. Proustite reaches is highest development in Chile, at the Dolores Tercera silver mine. The pictured 7-cm specimen is one of the finest surviving examples of the prismatic habit, showing numerous large, gemmy, lustrous, undamaged crystals.

58 PYROMORPHITE

BUNKER HILL MINE, KELLOGG, COEUR D'ALENE REGION, SHOSHONE COUNTY, IDAHO The 11.6-cm pyromorphite cluster pictured here is among the best examples recovered from the 1980 discovery at the Bunker Hill mine. The rich green and yellow colors, the large crystal size, the high luster, the overall degree of perfection, and the aesthetic way in which the crystals are arranged make this a connoisseur's specimen.

59 QUARTZ *variety* AMETHYST

ARTIGAS, ARTIGAS DEPARTMENT, URUGUAY Large amethyst-lined geodes have been mined near Artigas in northern Uruguay for decades. The crystals typically line open bubble-shaped voids in volcanic rock and sometimes occur as crystal-covered stalactites. The 15.2-cm stalactitic example shown here has fine color, excellent luster, and a high degree of perfection.

60 QUARTZ (SCEPTER)

BERILO BLANCO MINE NEAR SAPUCAIA DO NORTE, MINAS GERAIS, BRAZIL Some scepter quartz specimens are formed when a second generation of quartz begins to crystallize on the tip of an earlier-formed crystal. No one is sure why the newer crystal is the only one of the two to grow from that point onward, but eventually it becomes wider than the host crystal underneath, and the combination resembles a royal scepter. The 25-cm example pictured here is unusually large and very perfect, consisting of a colorless host crystal and a smoky scepter top. The famous locality (known for the world's finest pink quartz) also lends cachet to the specimen.

61 QUARTZ

with SIDERITE

MORRO VELHO MINE, NOVA LIMA COUNTY, MINAS GERAIS, BRAZIL The Morro Velho mine is the source of what many believe to be the most attractive quartz crystals found in Brazil – a significant statement inasmuch as Brazil has a great many quartz localities. The splendid 12.8-cm specimen pictured also includes attractive siderite crystals and pyrrhotite crystals as the matrix for the large, undamaged quartz crystal.

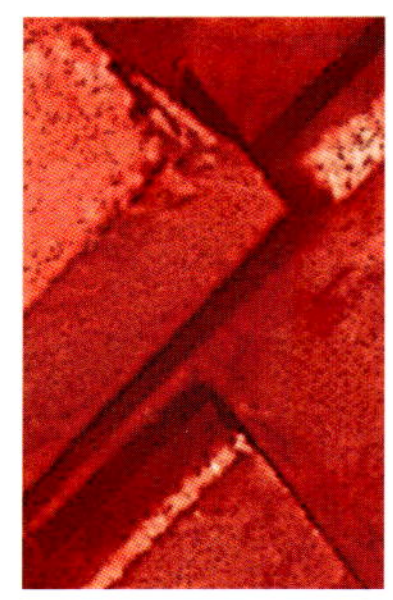

62 RHODOCHROSITE

(NAMED "THE ALMA QUEEN")

SWEET HOME MINE NEAR ALMA, COLORADO The superb 10-cm rhodochrosite crystal on matrix pictured here, known as "The Alma Queen," is among the most famous specimens in all of mineralogy. It is the best specimen recovered from a pocket found in 1965 by John Soules and Warren Good. The exceptionally large, rich red-colored crystals perched on a crystallized matrix of quartz and tetrahedrite combine to create a level of sculptural composition and aesthetic beauty that make this one of the world's truly great mineral specimens. Since its discovery, it has attained legendary status as a "Mona Lisa" of the mineral world – it is widely regarded as the finest example of rhodochrosite known – and among knowledgeable collectors and connoisseurs, it is considered to be the finest mineral specimen of any kind ever found.

63 RHODOCHROSITE

SWEET HOME MINE NEAR ALMA, COLORADO A group headed by Bryan Lees of Golden, Colorado, reopened the Sweet Home mine in 1991 and have since recovered many truly extraordinary specimens. Their greatest discovery was "The Good Luck Pocket," opened September 21, 1992. From that crystal pocket came many incredible specimens, including the one pictured here, the finest piece recovered from the mine in nearly 30 years.

64 RHODOCHROSITE

N'CHWANING MINE, KURUMAN, NORTHERN CAPE PROVINCE, SOUTH AFRICA The epitome of South African rhodochrosite is represented by gemmy, deep red scalenohedral crystals in solid clusters such as the beautiful 9.5-cm example pictured here. Though not quite as spectacular as the big rhombohedrons from the Sweet Home mine in Colorado (see plates 62 and 63), these clusters are highly valued. Their deep red color, high transparency, brilliant sparkling luster, sharp crystal form, and large, aesthetic clusters are second only to the Colorado specimens.

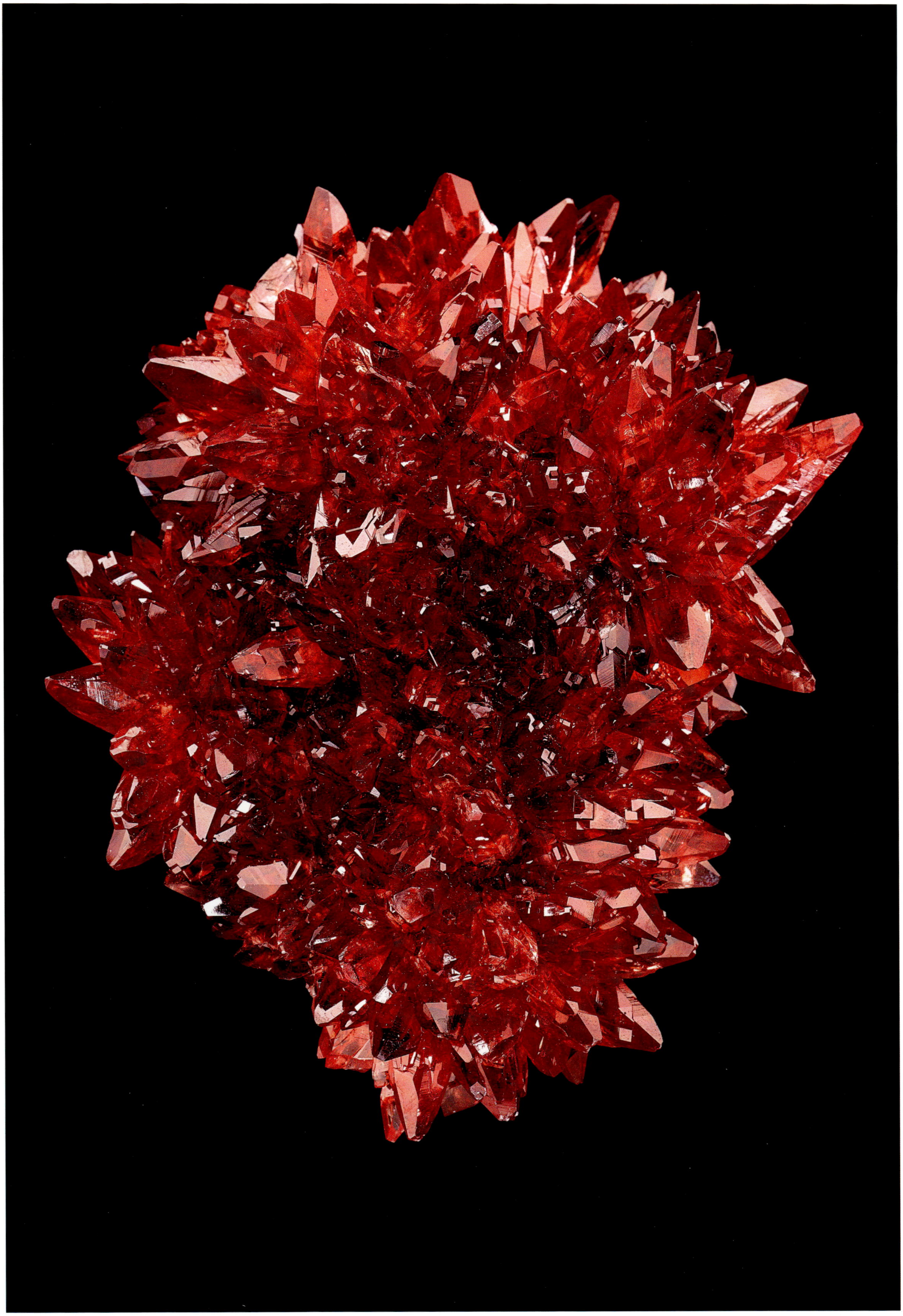

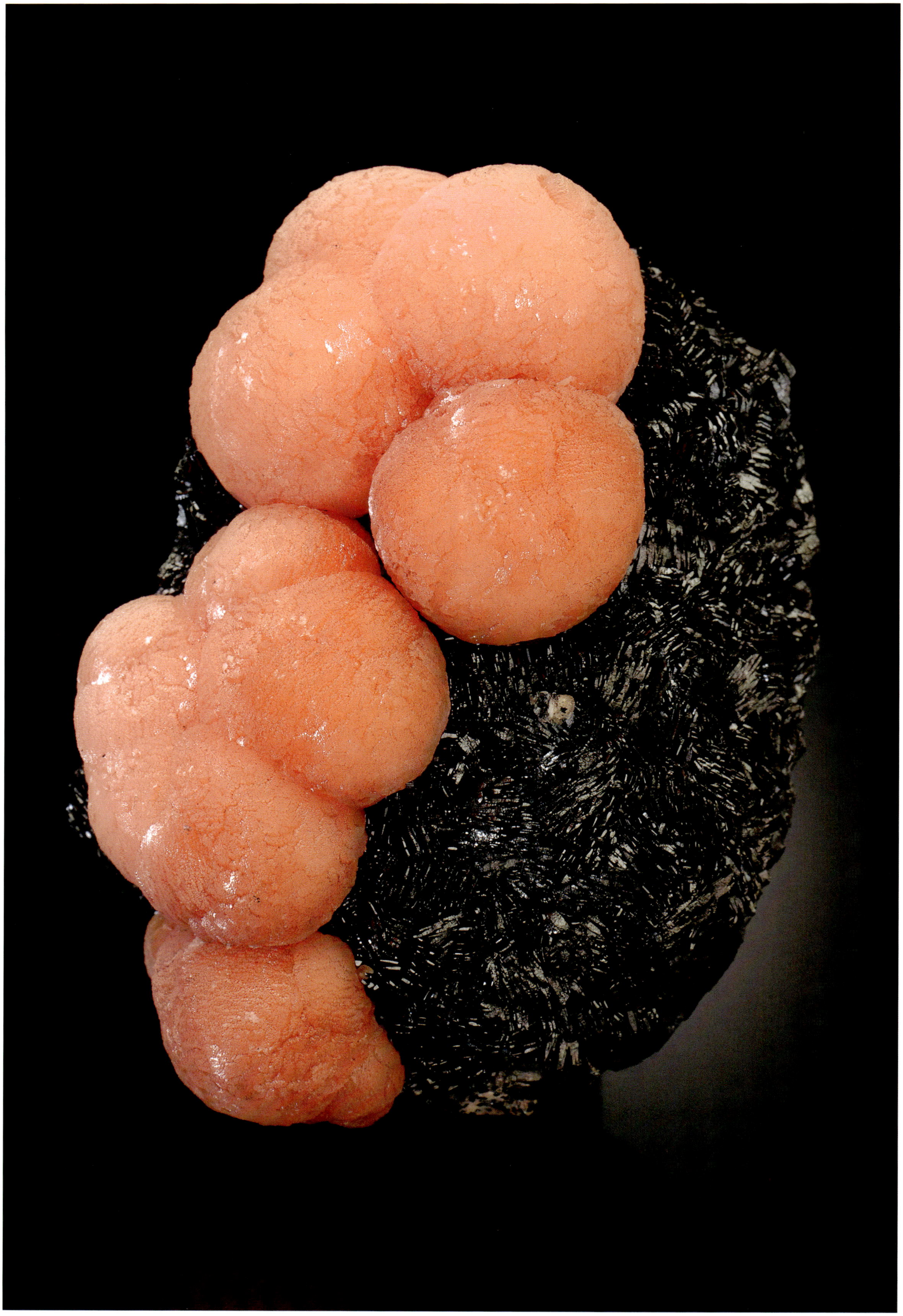

65 RHODOCHROSITE *on* MANGANITE

N'CHWANING MINE, KURUMAN, NORTHERN CAPE PROVINCE, SOUTH AFRICA The 5.2-cm specimen of rhodochrosite spheres on black manganite matrix pictured here represents a class of specimens unique to the Kuruman deposits. The rhodochrosite is not gemmy or darkly colored but is so interesting in a sculptural sense that the specimens are much admired by collectors.

66 RHODOCHROSITE

UCHUCCHACUA MINE, OYON PROVINCE, LIMA DEPARTMENT, PERU This beautiful, 4.5-cm crystal of rhodochrosite is the largest known crystal from the Uchucchacua mine. Though such specimens are not quite in the same league as rhodochrosites from the Sweet Home (see plates 62 and 63) and N'Chwaning (see plates 64 and 65) mines, the Uchucchacua mine is still one of the premier rhodochrosite localities in the world. This locality, high in the Peruvian Andes Mountains, has produced fine specimens since the early 1980s.

67 SCORODITE

EL COBRE MINE, ARANZAZU, MUNICIPIO DE CONCEPCIÓN DEL ORO, ZACATECAS, MEXICO The world's finest crystals of scorodite were found in a single pocket that yielded about a dozen specimens in 1968. The crystals are extraordinarily sharp, lustrous, well formed, and undamaged, and they rest on a contrasting matrix of white quartz. Furthermore they exhibit the "alexandrite effect," appearing to be purple in incandescent light and turquoise-blue in daylight. The 6-cm specimen shown here is among the best from that find.

68 SILVER *with* CALCITE

KONGSBERG, NORWAY Wire silver from Kongsberg, Norway, has been a classic mineralogical collectible for centuries, and is highly sought after. The 10-cm specimen pictured here is particularly desirable for its sculptural appeal. Most of the largest wire silvers from Kongsberg are preserved today in the Norwegian Mining Museum in Kongsberg.

69 SILVER

with CALCITE

KONGSBERG, NORWAY The 13.2-cm specimen pictured here is particularly desirable because of the calcite crystals entangled in the wires, and especially because of the large and lustrous crystal of black sphalerite clinging to the uppermost wires.

70 SILVER

QUINCY MINE, PEWABIC LODE, HOUGHTON COUNTY, MICHIGAN Large clusters of branching silver crystals such as this 9-cm example from the Quincy mine are very rare among Michigan specimens, even though many smaller silver specimens have been saved. The sculptural, branching forms make it particularly interesting.

71 SMITHSONITE

TSUMEB MINE, TSUMEB, NAMIBIA Among the most sought-after of all smithsonite specimens from anywhere in the world are the very rare turquoise-blue crystals from Tsumeb. The color comes from trace amounts of copper; cuprian smithsonite, however, is usually green instead of blue, making the blue specimens especially rare. The extraordinary 7.5-cm example pictured here shows good color, brilliant luster, and attractive crystals in an aesthetic grouping.

72 SMITHSONITE

TSUMEB MINE, TSUMEB, NAMIBIA Among the most beautiful of all smithsonite specimens from anywhere in the world are the apple-green to emerald-green crystals from Tsumeb. The color comes from trace amounts of copper and is so distinctive that cuprian smithsonite from Tsumeb was, for a time, given its own name, "herrerite" (named after the local Herrero tribe) before it was proven to be a variety of smithsonite. The extraordinary 8.1-cm example pictured here is perhaps the finest known example of the variety, showing deep color, unusual transparency, brilliant luster, interesting crystal shape, and a fine, large grouping of crystals.

73 SPANGOLITE

CZAR SHAFT, COPPER QUEEN MINE, BISBEE, COCHISE COUNTY, ARIZONA The world's finest specimens of this rare copper mineral have come from the famous Bisbee copper mines. This piece, from the Czar shaft, with crystals to 1.5 cm on a bed of green malachite crystals, is the finest known. Because of its extreme rarity, beauty, and composition, many connoisseurs consider this to be among the world's great mineral specimens.

74 SPODUMENE

variety KUNZITE

PALA CHIEF MINE, SAN DIEGO COUNTY, CALIFORNIA When the mineral spodumene occurs in lilac-colored crystals it is known as the variety *kunzite* (after the great American gem expert George F. Kunz, 1856–1932). The gorgeous 30-cm (1 foot!) crystal shown here, from the Pala Chief mine in San Diego County, California, is one of the finest North American kunzite crystals ever recovered and the best that has been found in California.

75 SULFUR

COZZODISI MINE, AGRIGENTO PROVINCE, SICILY, ITALY Sicily was famous for its sulfur mines throughout the late 1800s and into the 20th century. Mining was conducted by relatively primitive methods, which allowed miners to stop and pick up extraordinary examples like the large and perfect 5.5-cm crystal on matrix pictured here.

76 TOPAZ

ALABASHKA, ALABASHKA-MURSINKA PEGMATITE FIELD, YEKATERINBURG OBLAST, RUSSIA The 4.3-cm specimen shown here, undoubtedly a very old one, exhibits the typical form of Alabashka topaz: a stout, blocky shape with flat, lustrous faces and good transparency. It is still attached to its host matrix of muscovite and albite.

77 TOPAZ

ANTÔNIO PEREIRA MINE, ANTÔNIO PEREIRA, DOM BOSCO, MINAS GERAIS, BRAZIL Brazilian "imperial" topaz having a golden orange to reddish-orange color has been a coveted gemstone since first discovered in the 18th century. The 12-cm crystal pictured here, found in the late 1970s, is among the largest and best to be recovered. Its high transparency, lack of internal flaws, and deep color give it exceptional value as faceting material, a factor that causes most such large crystals to be cut rather than saved as specimens.

78 TORBERNITE

MUSONOI MINE NEAR KOLWEZI, KATANGA, DEMOCRATIC REPUBLIC OF THE CONGO Pockets in uranium-bearing ore lined with gorgeous, dark emerald-green clusters of thin square crystals of the copper-uranium phosphate torbernite have been highly valued by mineral collectors since the 1950s, when great specimens like the 9-cm example pictured here were discovered at the Musonoi mine. Although the mineral has been found at various neighboring mines, the world's finest torbernite crystals and crystal clusters have come from the Musonoi mine.

The same tanzanite viewed from three angles exhibits entirely different colors.

79 ZOISITE *variety* TANZANITE

MERELANI HILLS NEAR ARUSHA, UMBA VALLEY, TANZANIA This magnificent, near-flawless crystal of tanzanite (the gemstone variety of the mineral zoisite) was found in Tanzania in 1991. As the accompanying illustration demonstrates, it is an exceptional example of the property of trichroism, whereby the same crystal exhibits three different colors depending on the viewing angle.

80 ZOISITE *variety* TANZANITE

MERELANI HILLS, SIMANJIRO DISTRICT, LELATEMA MOUNTAINS, ARUSHA PROVINCE, TANZANIA The superb crystal of tanzanite (blue zoisite) pictured here measures 10 cm and was collected in 2003. As with most large stones, this one has great value for faceting. However, its excellent luster, well-formed faces, and freedom from damage make it a crystal worth preserving. The gem value helps to make it a true connoisseur's piece.

THE PATHS OF DISCOVERY

From the Mine to the Museum

ACANTHITE

LAS CHISPAS MINE, ARIZPE, MUNICIPIO DE ARIZPE, SONORA, MEXICO

The Las Chispas mine, also known as the Pedrazzini mine, was a silver and gold operation exploiting the Las Chispas vein system. The mine was opened sometime in the 1880s by the Santa Maria Mining Company, but the fledgling operation failed to turn a profit because the miners kept stealing the attractive silver minerals. In 1907 the mine's clerk, John Pedrazzini, was permitted to take over ownership of the property in lieu of back pay, provided that he also pay the back taxes that were due. He formed the Pedrazzini Gold and Silver Mining Company and found such an abundance of sparkling, crystallized silver minerals that he named the mine *Las Chispas*, "The Sparks."

Pedrazzini ultimately uncovered a system of 14 parallel fissure veins containing extremely rich ore averaging an incredible 500 ounces of silver and 4 ounces of gold per ton. In 1970 more than 160 pounds of acanthite crystals were recovered from the veins and were smelted for their silver content, but the miners stole ("highgraded") many additional specimens, which they sold in secret, probably including the specimen pictured here. Las Chispas has also produced fine crystal specimens of other silver minerals including stephanite, pearceite, native silver, and polybasite.

LEFT: *Stock certificate for the Pedrazzini mines, Arizpe, Sonora, Mexico, 1924.*

ADAMITE

SAN JUDAS CHIMNEY, LEVEL 6, OJUELA MINE, MAPIMI, DURANGO, MEXICO

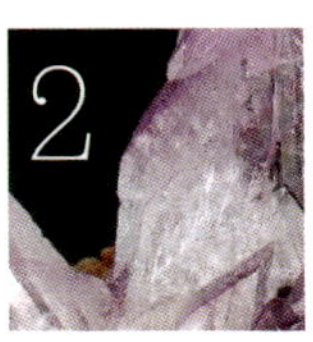

The Ojuela district, widely considered Mexico's greatest mineral locality, was discovered by the Spanish in 1598. It was not until 1945, however, that adamite was first found there, and significant numbers of specimens began to emerge on the market.

The orebodies at Ojuela comprise an enormous, interconnected series of chimneys and pipe-like deposits extending to a depth of more than 900 meters. The purple adamite pocket was encountered in 1981 by specimen miners at a depth of 600 to 700 meters in an area that was formerly below the water table and flooded. The best specimens were removed by unauthorized "highgraders" who plied the guard with alcohol , but were later bought back by the specimen-mining consortium Artrox.

Thus far, 117 different mineral species have been found in the Ojuela mine, including the world's best adamite. (The pictured specimen is the finest known example.) Green, yellow-green, white, and turquoise-colored adamite have also been found there.

LEFT: *Anchoring towers of the Roebling suspension bridge at the Ojuela mine, ca. 1893.*

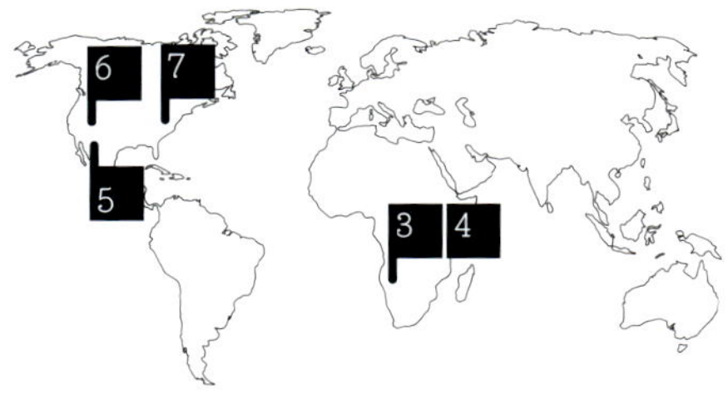

AZURITE

TSUMEB MINE, TSUMEB, NAMIBIA

The Tsumeb ore pipe was discovered in a remote region of German South-West Africa (now Namibia) in 1893, appearing as a bright green ridge rising above the desert. The green color came from the rich copper mineralization. Mining there began in 1907, and many extraordinary specimens of a variety of mineral species were recovered as mining passed down through the near-surface oxidized zone. By the time the oxidized zone had been penetrated and mining entered the primary sulfide zone, the mine was a world-famous mineral locality. Because good specimens usually come only from the oxidized zone, it was thought that Tsumeb's heyday was more or less over. However, due to a geological peculiarity, a *second* oxidized zone was encountered in 1957 on the 28 level, at a depth of 1,800 feet, and a whole new crop of extraordinary specimens was uncovered. In recent years a third oxidation zone has been found on the 44 level.

All of the great azurite specimens from Tsumeb, such as these, were found in the upper oxidized zone of the orebody. However, azurite specimens have been recovered throughout the mine's history because, as the lower levels were developed, mining continued on a smaller scale in the upper levels to provide necessary oxidized zone material for the mill and smelter feed.

The Tsumeb ore pipe eventually narrowed down with depth, and the mine became unprofitable to operate; it was closed a few years ago. Tsumeb is now close to being considered an "extinct locality" due to abandonment and flooding.

LEFT: *The so-called* Himmelsleiter *("Ladder to Heaven") hoisting ore cars out of the Tsumeb mine, ca. 1924.*

AZURITE

HOLBROOK MINE, BISBEE, COCHISE COUNTY, ARIZONA

The history of Bisbee is one of the world's great copper mining sagas. Mineralization in the area was first discovered in 1877 by John Dunn, a member of a military scouting party from Camp Bowie. Dunn told George Warren about his find and financed or "grubstaked" him. Warren staked many claims but did not include Dunn as a partner. As karma and legend would have it, Warren lost his holdings in a footrace against a horse and died in 1895, a penniless man. Nevertheless, his image is still part of the State Seal of Arizona.

By the mid-1970s, when the mines shut down, the district had produced nearly eight billion pounds of copper and more lead, silver, and gold than any other mining camp in Arizona. For over 100 years, this mining district continually produced magnificent specimens of azurite in a range of habits from flattened pseudo-rhombohedra to rosettes to botryoidal groups such as this fine specimen from the Holbrook mine.

The Holbrook shaft was sunk in 1889 by the Holbrook and Cave Development Company, a group owned by families that controlled the famous Phelps Dodge Corporation. The property was transferred to the Copper Queen Company in 1892. The Holbrook mine operated primarily in oxidized ores and was one of the district's greatest producers of mineral specimens, particularly botryoidal azurite like the fine example pictured. The mine was swallowed by the growth of the Lavender open pit mine in 1969.

LEFT: *John Dunn, discoverer of the first copper ore, at present-day Bisbee, 1877.*

BARITE

MEICKLE MINE, ELKO COUNTY, NEVADA

Situated in the Carlin Trend gold belt in northeastern Nevada, the Meickle mine exploits localized high-grade gold orebodies. In the course of underground mining, begun in 1996, large caverns were encountered and some of these proved to be lined with beautiful barite crystals. The extraordinary crystal cavern found on the 328-meter level (source of the illustrated specimen) measured 49 meters (160 feet) in height and width. It was named the George Bush Vug, in honor of its most famous visitor, former U.S. President George H. W. Bush. Another crystal cavern of similar size was discovered on the 373-meter level, this one lined with beautiful barite and calcite crystals interspersed with stalactites and draperies. Some of the finest barite crystal clusters in the world have come from these caverns.

Many extraordinary specimens were recovered in a short time but, because the Meickle mine is a working mine, operations soon progressed through the caverns to lower levels that are relatively free of caverns and of barite mineralization.

BERYL *variety emerald*

NEAR HIDDENITE, ALEXANDER COUNTY, NORTH CAROLINA

Recovered in December 2003 by James K. Hill of Hiddenite, North Carolina, this specimen epitomizes the qualities of a truly exceptional, connoisseur's mineral specimen.

Factors such as intensity of color and the presence of large areas of fine-quality cutting material testify to the importance of the crystal as a gem-quality emerald. It is a singular mineral specimen because it has exceptional overall beauty that is derived from the perfection of the emerald crystal, the luster of its crystal faces, and, perhaps most important, the fact that emerald crystal remains attached to the original matrix of crystallized, calcite, rutile, and muscovite upon which it formed.

Originally discovered in the 1870s, and worked intermittently ever since, the emerald mines at Hiddenite have yielded numerous crystals up to several hun-

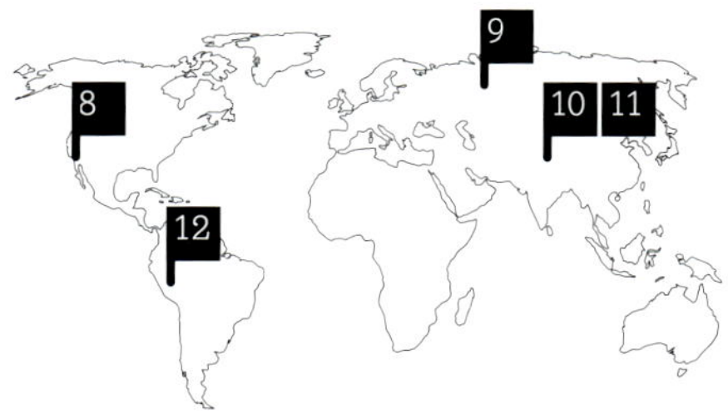

dred carats, although much smaller crystals are the norm. Most commonly, the crystals are light green, heavily flawed, and highly etched but several recent discoveries are the exception. Mr. Hill, President of North American Emerald Mines, first discovered high-quality emeralds at this locality in late 1998 and recovered several notable crystals in the ensuing years, which enabled him to begin full-scale mining operations in November 2003.

Within a week, a large intrusion of quartz was encountered several feet below the surface, convincing the miners that they were on the right track because similar rocks had been associated with previous emerald finds. Their intuition paid off when, in mid-December, miners broke into a cavity measuring 12 feet long and 5 feet wide. Astonishingly, the walls of the cavity were lined with countless crystals of quartz, pyrite, muscovite, rutile, and calcite – and a single crystal of emerald that set a new standard for quality.

BERYL *variety morganite*

WHITE QUEEN MINE, SAN DIEGO COUNTY, CALIFORNIA

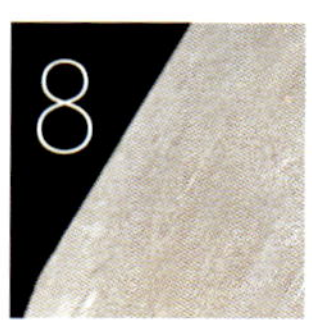

The White Queen mine is located on Hiriart Mountain in the Pala district, San Diego County, California. The pegmatite was apparently discovered around 1900 by Fred Sickler, but little work was done there until the 1950s. George Ashley had purchased the property from Sickler in 1947, then sold it to Norm Dawson in 1948. Dawson began tunneling into the pegmatite in 1959, and encountered a crystal-lined pocket large enough for a man to stand in. The pocket yielded 5,600 kg of large quartz crystals and numerous fine specimens of pink beryl totaling about 100 kg (220 pounds), much of which was sold for gem rough. Another large pocket, found in 1964, yielded 6,000 kg of quartz crystals and about 30 kg of pink beryl, including the pictured specimen. Yet another large pocket was opened by Dawson's sons in 1990; it yielded around 200 well-formed specimens of pink beryl in crystals to 8 cm, many on a matrix of white albite. Since that time, ownership of the property has reverted to the Pala Indian tribe, and little if any further work has been done.

BERYL *variety heliodor*

MURSINKA, ALABASHKA-MURSINKA PEGMATITE FIELD,
URAL MOUNTAINS, YEKATERINBERG OBLAST, RUSSIA

The Alabashka-Mursinka pegmatite field is the most famous granitic pegmatite region in all of Russia and the former Soviet Union, being most noted for superb gem crystals of beryl, topaz, and tourmaline. Yellow beryl crystals, such as this one (the gem variety "heliodor"), were first reported from the Mursinka pegmatites in 1837. In the years before 1886 visitors saw 44 pit mines around Mursinka in operation, but shortly thereafter only a few amethyst pits were still being worked.

Mining was resumed sporadically after the 1880s but came to a halt again following World War I, because the new Communist government wished to focus the people's labor on more practical production than items for personal adornment.

The Mursinka mines were rich in vugs and pockets filled with clay and gem crystals. Nikolai Koksharov, a Russian mineralogist of the time, commented that the Mursinka beryls (typically wine-yellow to greenish yellow in color) were the finest in all of the Urals.

BERYL *variety aquamarine*

DUSSO, NORTHERN AREAS, PAKISTAN

In 1912, gem-bearing pegmatites were discovered by local villagers in the cliffs overlooking the town of Dusso, Pakistan. Dusso is situated on the Braldu River in Gilgit Division, in the shadow of the lofty Karakoram Range, in northernmost Pakistan, near the borders with India, China, and Afghanistan. It is a remote and primitive area, very dangerous for outsiders, especially for those who cannot speak the various local languages (Balti, Shina, Khowar, Burushaski, and Wakhi). (This part of Pakistan has been in contention between Hindu and Muslim forces since 1947.) The Dusso pegmatites are rather simple in their composition, with vugs and pockets containing well-formed crystals of aquamarine, the finest of which are illustrated here.

BERYL *variety goshenite on microcline*

MINAS GERAIS, BRAZIL

The varietal name "goshenite" refers to crystals of colorless beryl and is derived from Goshen, Massachusetts, where colorless beryls were mined in the 19th century. While well-formed beryl crystals commonly occur in a wide variety of colors including blue (aquamarine), yellow (heliodor), pink (morganite), grass green (emerald), and red (red beryl), truly colorless crystals of beryl are quite rare. This example is particularly striking because the goshenite crystal is perched atop an equally well-formed crystal of microcline. The reddish-brown hue of the goshenite crystal and the light reddish-brown coloration on the surface of the microcline are caused by the presence of iron oxides.

The world famous gem producing regions located in the Brazilian state of Minas Gerais first became known to European explorers in the 1500s. Hearing native reports of green gemstones found inland and thinking that an emerald deposit must be hidden in the jungle, Portuguese pioneers mounted four expeditions up the major river valleys to search for gems, in 1554, 1568, 1571, and 1674. The first expedition, led by Francisco Spinoza, found no emeralds but his men

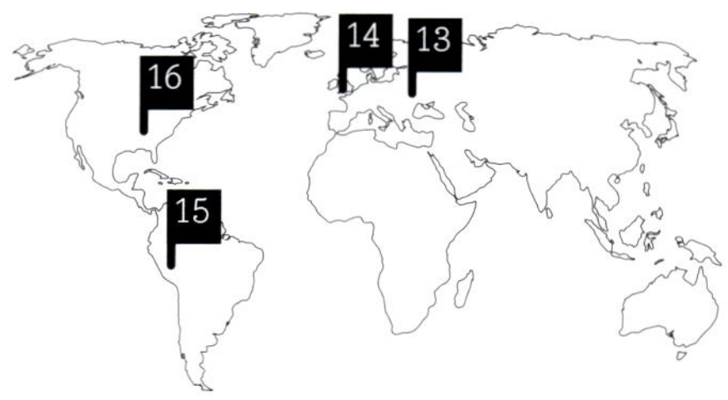

surely brought back crystals of green elbaite tourmaline and blue-green aquamarine beryl. A green Brazilian tourmaline was among the first pictures of any crystal to be published in a book. It appeared in a mineralogical anthology published by Conrad Gesner in Zurich in 1565.

BERYL *variety heliodor*

VOLODARSK VOLYNSK PEGMATITES, ZHITOMIRSKAYA OBLAST, UKRAINE

Large, yellow-green beryl crystals (and their heat-treated aquamarine equivalents) from Volodarsk Volynsk first reached the Western specimen market in 1978. In 1992, underground mining operations at Volodarsk Volynsk encountered beryl-bearing pegmatite at a depth of about 70 meters. The pegmatite body measured about 50 x 70 meters in plan and encapsulated a cavity zone 4 x 5 meters across and 12 meters long, which yielded more than a ton of collectible and gem beryl crystals, including the one illustrated here. Among the most distinctive features of these crystals is the corroded surface, which developed during the evolution of the pegmatite pocket, as surrounding fluids changed. Termination faces are deeply etched, showing small, irregular pyramid points and cavernous depressions; the crystal edges have become rounded; and the long sides of the crystals show a tile-like surface of rectangular hillocks resulting from dissolution – all of these surface features being very bright and lustrous. The Volodarsk Volynsk mines closed in 1995.

BOURNONITE

HERODSFOOT MINE, LISKEARD, CORNWALL, ENGLAND

Lead-silver lodes were discovered in Cornwall south of Liskeard in the 1840s. Ore output was never very large, but some of the mines became famous for their mineral specimens. During the years from 1858 to 1868 the Herodsfoot lead mine encountered sporadic quartz-lined cavities containing what are still to this day the finest crystals of bournonite ever found anywhere. The best specimens, including the one pictured here, were handled by the British mineral dealer Richard Talling, who lived 6 or 7 miles away in the town of Lostwithiel.

Today the brick chimney associated with the mine still stands, but the old ruins have been engulfed by forest. The fine specimens saved by Talling (who was once banned from the mine because of his activities but gained legal access again by buying one share of stock in the mining company) have continued to circulate among museums and major private collections, and are still treasured as the finest examples of the species.

BRAZILIANITE *with muscovite*

CÓRREGO FRIO NEAR LINOPOLIS, MINAS GERAIS, BRAZIL

The Córrego Frio pegmatite was discovered in 1942 by Alfredo Severino da Silva while cleaning up his cornfield on the banks of Cold Creek *(Córrego Frio)*. He gathered about 3 kg of crystals of a then-unknown but obviously valuable mineral from a weathered outcrop and showed them around, thinking perhaps they were chrysoberyl. Three other men from nearby Conselheiro Pena staked a claim and began trenching and digging at the outcrop, recovering another 40 kg of crystals. Disappointed that the crystals were softer and more cleavable than the more typical gem minerals, the miners abandoned the workings a short time later. However, specimens sold in Belo Horizonte came to the notice of American mineralogists Frederick Pough and Edward Henderson, who described the mineral as a new species in 1944, naming it after its country of origin.

Shortly thereafter the American mineral dealer/collectors Ed Swoboda and Fred Heuberg obtained a lease on the Córrego Frio property and began to supervise new excavations. During Swoboda's work there in 1945 and 1946 he recovered nearly all of the large and beautiful specimens that now grace museums and private collections around the world, like the example illustrated here. Landslides have since covered the excavations and the adits have caved in. In the 60 years since then no finer specimens have been discovered anywhere.

CALCITE *on barite*

ELMWOOD MINE NEAR CARTHAGE, SMITH COUNTY, TENNESSEE

In 1964 the New Jersey Zinc Company made the decision to begin exploratory drilling in the limestones along a geologic structure called the Nashville Arch in Tennessee. The geological conditions were similar to those surrounding well-known lead-zinc-barite-fluorite deposits in other lower Ordovician-age limestones, and the company hoped that those indicators were pointing to as-yet undiscovered deposits in Tennessee as well. In February of 1967 their hunch proved correct when drill hole number 79 penetrated 5 feet of rich zinc ore at a depth of 1,350 feet. Years of additional drilling were needed to establish the total extent of the orebody, but by August 1969 the new Elmwood mine was ready to go into full production.

Because one of the indicators for ore was karst topography (the prevalence of underground limestone caverns), it was not unusual for the miners to break into open caverns in the ore zone. These caverns were often rich in ore minerals, but what was particularly interesting was that the partial collapse of some of the caverns had resulted in huge piles of broken rock covering the cavern

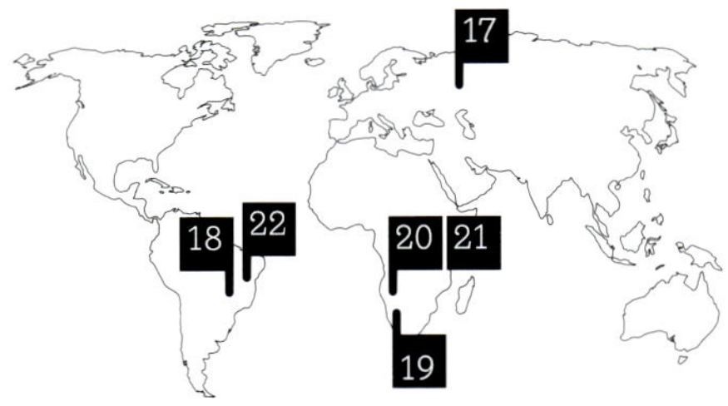

floors, and the loose pieces of rock commonly carried exquisitely fine crystals of fluorite, calcite, and sphalerite that had formed on them following the collapse. Thousands of specimens were recovered by miners and this piece is among the finest. Since then, however, most of the mines have ceased operations.

CALCITE

SARBAYSKAYA QUARRY, RUDNIY, KUSTANAY OBLAST, KAZAKHSTAN

Kazakhstan, formerly a tightly closed country, has been producing specimens for the international mineral market since the fall of the Soviet Union in 1989. Fine specimens of native silver, various metallic sulfides including bornite and pyrrhotite, dioptase, superb cassiterite crystals, and by far the world's finest purple creedite have come from its mines in the last 15 years. Normally the mineral calcite would not fetch sufficient prices to make it worth exporting from such a difficult and remote area. However, the abundance and beauty of the Rudniy crystals convinced mineral dealers to make the effort, preserving exceptional crystals like this one because of their high quality.

CALCITE

IRAI MINE, IRAI VILLAGE/STATE, RIO GRANDE DO SUL, BRAZIL

The Irai mine has produced large quantities of gem-grade and specimen-grade amethyst in various habits and sizes (some of which is heat-treated to produce a beautiful orange-yellow citrine), and also many attractive specimens of calcite in tapered white crystals on matrix like this one. The minerals are found lining large gas-bubble pockets in basaltic lava flows. In 2003 the locality was producing about 300 kg of carefully hand-cobbed and trimmed specimens per month, but the outlook was not promising because the workings were getting too deep to be extended safely. The Irai mine is located within an area occupied by indigenous people of the Kaingang tribe who are seeking protection of their rights and a degree of autonomy from the Brazilian government.

CERUSSITE

KOMBAT MINE, OTAVI, NAMIBIA

Minerals were first reported from the Kombat area in 1851, but because of its remote location the deposit lay undisturbed for the next 60 years. In 1911 the Otavi Mining and Railroad Company began mining there and recovered nearly 5,000 tons of ore before flooding caused the mine to shut down in 1925. The Tsumeb Corporation began exploratory drilling in 1954, locating additional orebodies, and began mining operations again in 1962. Mining proceeded profitably until 1988 when an underground aquifer (water-bearing formation) was struck, flooding the mine permanently. The pictured specimen was recovered in 1977 from a pillow-sized pocket that yielded a handful of incredibly brilliant, heart-shaped crystal twins that have never been surpassed at any other locality.

Kombat mine management was notoriously hostile to the preservation of specimens. Nevertheless, during the 1970s and 1980s a limited number of fine specimens of cerussite, azurite, malachite, and cuprite were preserved by miners. Several species new to science (including one that has since been named kombatite) and also the world's finest crystals of the gemmy red mineral nambulite were also recovered.

CERUSSITE

TSUMEB MINE, TSUMEB, NAMIBIA

Cerussite specimens have always been limited to the oxide zones; in the upper oxide zone it was found primarily in the workings between 150 and 250 meters in depth, and not at all below the 400-meter level until the second oxide zone was reached, and then again in a small third oxide zone. Crystals up to 60 cm in length have been reported. Inclusions or overgrowths of other minerals can turn cerussite a bright red (cuprite), deep green (malachite), gray (galena), pale green (smithsonite), yellow, and black. Some individual crystals are transparent enough to be faceted into gemstones. (See entries 3 and 4 for history of Tsumeb mine.)

LEFT: *Miners working in the original opencast at Tsumeb, ca. 1910.*

CHRYSOBERYL *Sixling Twin*

ITAGUAÇU, ESPIRITO SANTO, BRAZIL

Around 1924 two magnificently twinned crystals of yellow-green chrysoberyl were found in Espirito Santo State, Brazil. The local digger who found them sold them to a German gem supplier named Otto Richard who sent them on to Idar-Oberstein in Germany. They were purchased there by a New York gem dealer named Lucien Zell, who returned with them to New York and sold them to the American mineral collector Washington A. Roebling (1837–1926). William F. Foshag, a mineralogist at the Smithsonian Institution, knew of the two Roebling specimens and very much wanted to see more. So when Foshag's friend Allan Caplan was planning a buying trip to Brazil in 1940, Foshag asked him to check out the Colatina locality.

At Colatina, Caplan learned from an old man that a hotel owner in the town of Santa Tereza, 25 miles to the south, had some of the chrysoberyl specimens.

And indeed he did, a cigar box full. The hotel owner informed Caplan that the specimens had come from a small, isolated town located another 25 miles farther down the road, called Itaguaçu. Upon reaching that town, Caplan was taken to a small white cottage where he was greeted by a woman and her child (all of the local people were descendants of Polish immigrants from long ago and spoke a strange dialect of Portuguese, which no one else understood). She led him behind the cottage to a small hillside with a hole no more that 5 feet square – the source of the world's greatest chrysoberyl specimens. Caplan dug out many more crystals with his bare hands, purchased a few more the woman had, and hiked down the road to purchase a big one from a neighbor, finally accumulating half a grain sack full of crystals, which he took back to the United States. The 6.5-cm specimen pictured here is surely one of Caplan's, because little was recovered from the locality thereafter.

COPPER

KEWEENAW PENINSULA, HOUGHTON COUNTY, MICHIGAN

Native copper on the Upper Peninsula of Michigan was well known among the local Indians for thousands of years before it first came to the notice of Europeans. The French explorer Samuel de Champlain (1567–1635), founder of the City of Quebec, was presented with a 30-cm copper specimen by friendly Algonquin Indians who had collected it from the bank of the Ontonagon River. Other explorers made note of the copper resources, and eventually the government took serious notice. The first mining leases were granted by the government in 1843, and by the 1850s a copper mining section was active on the Boston Stock Exchange. Cornish mining men were brought in to oversee operations at the dozens of mines that sprang up, and a copper boom lasted for many years. Production from the peninsula had peaked by the time of World War I, and in the following years the mines gradually closed, one by one. Over 13 billion pounds of copper and 16 million ounces of silver were produced since 1845. Today only one mine remains open, but the legacy of copper mining still colors the entire region and draws historically minded tourists every year.

Many of the copper mines produced excellent specimens of native copper and native silver, though usually with crystals under 1 cm. The older specimens, such as the ones pictured here, often do not carry specific mine names because they were initially saved as mantle-piece souvenirs by miners rather than as geological or mineralogical specimens.

LEFT: *Miners with a large mass of native copper in the Quincy mine, ca. 1918.*

CORUNDUM *variety ruby*

MOGOK STONE TRACT, KATHA DISTRICT, MYANMAR (FORMERLY BURMA)

The Mogok Stone Tract in Myanmar (formerly Burma) has been the world's principal source of fine gem rubies for over 800 years. Legends of the rubies go back to ancient times, and in fact the rubies have probably been known there since the earliest humans arrived in the area. Nicolò de Conti (1419–1444) was among the first Europeans to visit the area, and described seeing the local king riding a white elephant draped in a gold chain set with many rubies. The British annexed the Mogok area in 1885, over commercial disputes and reports of atrocities committed by King Thebaw. The tract was put up for lease and eventually acquired by Burma Ruby Mines, Ltd. Some of the richest alluvial deposits were found directly beneath the town of Mogok itself, necessitating the relocation of citizens. Commercial mining in the jungle was more difficult than had been foreseen, though, and the company folded in 1931. In 1969 all mines were nationalized, and ruby smuggling increased. As of 1992 the government was operating seven ruby mines by both open pit and underground methods.

LEFT: *Baskets of ruby-rich ore being winched to the surface at Mogok, Burma.*

CORUNDUM *variety sapphire*

RATNAPURA, SRI LANKA

The island of Sri Lanka (formerly Ceylon) off the southern tip of India has been a source of colored gemstones since ancient times. The earliest document referring to gem mining is the *Rathu Kata,* describing ruby mining operations near Ratnapura over 2,500 years ago. Ratnapura is also the earliest known site for the mining of blue sapphires (called *nilkata*). Alexandrite, amethyst, garnet, topaz, sinhalite, taaffeite, spinel, tourmaline, zircon, and other colored gem species are also recovered on the island. Over 100 rivers drain the rainforests there, where for millions of years the underlying rocks have been slowly weathering away, freeing their precious cargo of gem crystals to be carried down the rivers and deposited with sand and gravel. These deposits are mined and sorted by hand, using primitive methods, at over 5,000 registered and 30,000 unregistered mining locations, mostly in the Ratnapura (meaning "Gem Area") and Elahara districts. From these modest operations Sri Lanka exports a quarter of a billion dollars in gemstones every year. Most of the gem crystals found are rather waterworn, but those recovered close to their source are the sharpest in form and the most desirable as specimens for the collector and connoisseur. The example shown here is remarkably sharp and unworn.

CROCOITE

ADELAIDE MINE, DUNDAS, TASMANIA, AUSTRALIA

Crocoite, the rare chromate of lead, has been found at a number of localities worldwide, but the finest and most sought-after specimens have always come from Tasmania. Crocoite was first reported there in 1885, in the Heazlewood mine. Silver ore had been found near Dundas in 1886 and by 1890 several mines were in production there, including the Adelaide mine. Ore yield was good to intermittent up to World War I, and crystals up to 3 inches were reported, but specimens were generally not saved, except for mantle pieces. Crystals up to 1 foot long were reported from nearby mines but none were saved; all went to the crusher and from there to the smelter. Commercial mining in the district has long since ceased.

Phoenicochroite, dundasite, and chromian yellow-green cerussite crystals were also found at the Adelaide mine. Large intergrown masses of crocoite to more than a foot across have been recovered by specimen mining operations in recent decades, but no specimens have been found to rival the simple pair pictured here.

CUPROSKLODOWSKITE

MUSONOI MINE NEAR KOLWEZI, KATANGA, DEMOCRATIC REPUBLIC OF THE CONGO

Uranium was first discovered in Katanga Province in 1913, at the Luiswishi copper mine, and subsequently also at a number of other cobalt-copper deposits. Uranium minerals were found at the Shinkolobwe copper mine in 1915, at the Kalongwe mine in 1930, and at the Musonoi mine in the 1950s. Only at Shinkolobwe were the uranium minerals plentiful enough to be mined as ore (Shinkolobwe uranium was used in the Hiroshima atomic bomb). At the other localities the amounts present were of little economic importance, although they spurred great mineralogical interest.

Cuprosklodowskite is the copper-rich analog of the uranium mineral sklodowskite (named after Marie Curie, whose maiden name was Sklodowska). It was first found at the Kalongwe copper mine, and later at various other mines in the district known as the Katanga Crescent, including the Kambove, Kasompi, Kolwezi, and Luiswishi mines. But only in the western sector of the Musonoi copper-cobalt mine, in a zone measuring 50 x 100 meters, has it been found in such large and beautiful specimens as this one.

DIAMOND

MIR PIPE, YAKUTIA, SIBERIA, RUSSIA

Rumors first arose in the early 1800s that precious stones could be found in Yakutia. Following the Communist Revolution in 1918, a Yakutia naturalist named P. K. Starovatov met two old local men who each reported finding what later proved to be diamonds in the sands of the Chona and Kempendyaika Rivers. In 1947 an organized government-sponsored search for diamonds began and in August of 1949 the first diamondiferous placer was found in a sandy spit in the Vilyuy River. By 1953 geological expeditions had identified a number of other diamond placers in Yakutia. The first diamondiferous kimberlite pipe was found the following year, and in 1955 a group of geologists located a particularly large and rich diamond pipe, which they named Mir, the Russian word for "peace." Exploration continued, and by 1959 no less that 120 kimberlite pipes had been discovered in Yakutia.

Because the Mir pipe was located closest to the transportation route down the Lena River, exploitation began there first. By the end of 1959 a total of 13,000 carats had been recovered and sold on the international market. The Mir pipe has been exploited by the open pit method, and the pit today is over three-quarters of a mile across and a quarter mile deep. Preparations are under way to convert the mine to an underground operation.

Since 1959, the Mir pipe has produced billions of dollars worth of diamond crystals of the finest gem quality. In December of 1981 this deposit produced a single diamond crystal weighing 342.5 carats, which was named "The 26th Congress of the Communist Party of the USSR Diamond." Numerous crystals have been recovered in the 200- to 350-carat range, including the 320.6-carat "Alexander Pushkin" diamond, which was recovered in 1989. By order of the Russian government, diamond crystals weighing more than 20 carats are preserved uncut and added to the National Treasury. The exceptional diamond pictured here, measuring four-fifths of an inch across (weighing approximately 11 carats), is particularly remarkable for having been preserved in its original matrix.

LEFT: *Heavy equipment in the Musonoi open pit, 1976.*

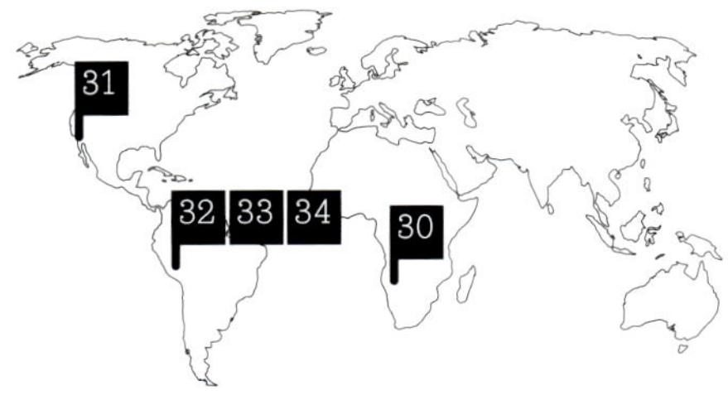

DIOPTASE

TSUMEB MINE, TSUMEB, NAMIBIA

Apparently dioptase was totally absent in the first oxidation zone, as no specimens were reported from that area before 1967, and none have been found in the old collections. The first Tsumeb dioptase specimens to reach the mineral market came out in 1975 and were collected between the 30 and 32 levels. Crystals from this zone were said to reach 5 cm in very rare cases, and enormous quantities were recovered in the late 1970s and early 1980s, almost to the exclusion of any other minerals. Much later a few specimens with crystals to 1 cm were found on the 44 level in the third oxide zone. The best of the Tsumeb dioptase still ranks as the finest known examples of the species. (See entries 3 and 4 for history of Tsumeb mine.)

ELBAITE *on quartz*

TOURMALINE QUEEN MINE, SAN DIEGO COUNTY, CALIFORNIA

The Tourmaline Queen mine, located on the northeast slope of Queen Mountain, is among the most well-known tourmaline-producing pegmatites in the world. It was originally staked as a quartz claim by Frank Salmons and others in 1903, and within a few years had become one of the leading sources of tourmaline in California. Many large crystals were shipped to the Dowager Empress of China, T'zu Hsi, for carving purposes. However, when the Imperial Chinese aristocracy was overthrown in 1912, the Chinese market for carving-grade tourmaline dried up, and the California mines closed.

Minor small-scale mining was carried out at the Tourmaline Queen mine in the 1940s and 1950s, but in 1971 a California company, Pala Properties, began full-time exploration and mining at the site. In January of 1972 an interconnected series of pockets found by William Larson and Edward Swoboda yielded the now-world-famous "blue cap" tourmaline specimens, of which the pictured example is the finest found. The specimens from this discovery are ranked as the finest tourmaline specimens ever found in North America.

ELBAITE

SALINAS MINE, ARAÇUAÍ-SALINAS DISTRICT, MINAS GERAIS, BRAZIL

The Araçuaí-Salinas pegmatite district is the richest tourmaline-producing area in all of Brazil. The Salinas pegmatite area, situated about 42 km northwest of the town of Coronel Murta, was discovered around 1885 by a Mr. Pego, and mining rights passed through a number of hands before much mining was undertaken. Around 1945–1950 a group headed by Telesco de Mattos had struck a gem pocket there, which yielded over a ton of tourmaline crystals. Mattos sold the

mine properties to Antonio Pinheiro, who greatly expanded the operation and blocked out the ever-increasing discoveries of the full extent of the pegmatite complex. In 1973 the mine was purchased by its present owner, Halley Batista, who has been the most successful of all. The Salinas mine has since become one of the world's most important sources of tourmaline in beautiful, gemmy crystals ranging from shades of pink and red (like the illustrated example here) through purple, green, blue, brown, and colorless.

Most of the Salinas pegmatite bodies have been completely altered, meaning that the white feldspars comprising much of the pegmatites have weathered to white clay, and the gem crystals are now floating loose in this material. Consequently mining proceeds with the use of a bulldozer, cutting downward gradually as each turnover of clay is carefully examined for crystals. The area of the pegmatites now contains mined canyons up to 30 meters deep and covers an area of 2,471 acres.

LEFT: *Open cut exploiting a gem-rich pegmatite body at Barra de Salinas.*

ELBAITE *on quartz*

PEDERNEIRA MINE, SAN JOSE DA SAFIRA, MINAS GERAIS, BRAZIL

The Pederneira mine is located near the hamlet of Santa Maria do Suaçuí, which is situated about 20 miles north-northwest of the large town of San Jose da Safira. It has operated on a parallel series of pegmatites since around 1980, producing primarily gem rough. Recently, however, a new pegmatite vein was opened about 100 meters uphill from the main mine; these new workings, called "Dada's Tunnel," began producing some extraordinary tourmaline specimens in 1999 (at the time labeled "Urupuca mine"), consisting of long green crystals surrounded by bright pinkish lavender lepidolite. This specimen was among the finest Pederneira mine tourmaline specimens that became available in February 2001 – with green, blue, and red crystals to around 8 or 10 cm or more, some in clusters on a matrix of white feldspar, purple lepidolite, and quartz.

ELBAITE

JONAS MINE, ITATIAIA, CONSELHEIRO PENA, MINAS GERAIS, BRAZIL

What is surely the world's greatest tourmaline discovery took place in 1978 near the small Brazilian village of Itatiaia, about 3 miles east of Conselheiro Pena in Minas Gerais State. A garimpiero (local itinerant miner) named Ailton Barbosa obtained the backing of a Governador Valadares gem dealer named Jonas Lima to reopen the João Pinto mine, a small pegmatite that had yielded multicolored tourmaline crystals up to 12 cm in the early 1940s. They drove an adit (tunnel) horizontally into the hillside and finally intersected the pegmatite. They first

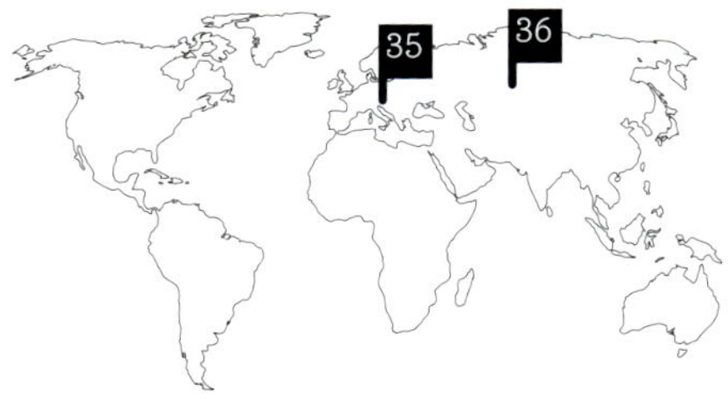

encountered pockets containing beautiful white albite crystals to 5 cm with pencil-thin green tourmaline crystals, then found several large pockets containing only worthless black tourmaline crystals with albite. Discouraged and on the verge of quitting after nearly six months of expensive and grueling work, they next broke into another worthless pocket of black tourmaline filled with mud and water. Ailton believed that yet another pocket must lie directly overhead, from which water was flowing into the lower pocket. On this hunch he had the miners carefully dig upward, breaching a small opening into another pocket. Ailton reached through the opening and withdrew a gorgeous red gem crystal of tourmaline. The pocket proved to be large enough for several men to stand in (about 8 feet wide and over 10 feet high), and was uniformly lined with clean, bright, gemmy, cranberry-red tourmaline crystals, some of enormous size laying on the pocket floor or suspended from the sides and ceiling of the chamber. The crystals ranged from small size up to a yard long, with large white albite crystals and quartz crystals. A pair of tourmaline crystals 50 cm and 30 cm in size and attached to the ceiling was cut loose and allowed to fall on a pile of old car tires to prevent damage; including its matrix the specimen weighs 775 pounds. Another large crystal called "The Rocket" weighs 300 pounds and measures over a meter in length. On the floor of the pocket, amidst countless smaller crystals, lay an 85-cm crystal weighing 180 pounds. Many crystals of all sizes were ultimately removed from the pocket during the following eight days, and have brought millions of dollars for Jonas Lima. The piece shown here is one of the most perfect specimens recovered.

EPIDOTE *on calcite*

KNAPPENWAND, UNTERSULZBACHTAL, AUSTRIA

The Alpine clefts of the Knappenwand occurrence were discovered in 1865 by a shoemaker, mountain guide, and mineral collector named Alois Wurnitsch. The first find consisted of crystals to 7 cm removed from a water-filled pocket. Wurnitsch took the specimens to his friend Andreas Bergmann, a tailor and part-time mineral dealer in Innsbruck, who sent them for identification to Professor Victor von Zepharovitch at the University of Prague. Zepharovitch was an excellent mineralogist who immediately recognized the specimens as the finest examples of epidote then known. While he carried out detailed research on the specimens (the locality for which was still being kept secret, as was customary), Andreas Bergmann obtained a collecting lease on the site from the Imperial Austrian Forest Service in 1867 and began digging for specimens. After three months of fruitless work in extremely hard rock, Bergmann broke into a new pocket and found specimens far surpassing the initial discovery. Other major pockets were to follow, collected with meticulous care by Bergmann and his miners, and marketed in person by Bergmann on his trips to museums and important collectors across Europe. Bergmann finally gave up the lease after several years and let others try their luck. Some of his successors, including Alois Hollaus in 1905, had the good fortune to find major pockets as well; Hollaus's pocket yielded nine knapsacks full of epidote specimens. Among the most successful of later collectors were Kajetan Stockmaier and Ehrenreich Schuchter, who carefully worked the mine from 1948 to 1956 and recovered many fine specimens that are among the best ever found.

LEFT: *Early lithograph depicting the Sulzbach Valleys in Austria. The Untersulzbachtal is the valley on the left.*

FERRO-AXINITE

PUIVA MINE, SUBPOLAR URAL MOUNTAINS, TYUMEN OBLAST, CHANTY-MANSIJSK AUTONOMOUS REGION, RUSSIA

Mineralization in the Puiva area was discovered in 1936 by government-sponsored prospecting expedition. During the years before World War II the extent of the deposit was determined, and approximately 100 tons of piezoelectric quartz crystals, recovered from the near surface zone, were shipped. Nothing more was done at the site for the next 20 years, but in 1961 it was reopened both as opencast and as underground workings. Once again, industrial-grade quartz was the focus of these operations, and by 1974 another 120 tons of crystals had been taken out. In 1974 vein masses of transparent quartz were also discovered, and from 1974 to 1987 over 2,500 tons of quartz were recovered. By the 1980s a depth of over 650 meters had been reached, and in 1984 work on the 705-meter level intersected 1 x 1.5-meter cleft, which yielded about 30 kg of mediocre ferro-axinite crystals. In 1990 a 9 x 15-meter cleft was discovered which yielded the world's finest crystals of ferro-axinite, up to 20 cm each, associated with fluorapatite and fluorapophyllite. The ferro-axinite proved to be so valuable as specimens that one section of the mine was turned over exclusively to specimen recovery. Over 50 cleft-like pockets have so far been encountered at depth. For size, luster, transparency, sharpness, and form, the Puiva ferro-axinites have no rival; the fact that many are recovered on matrix is even more remarkable considering the hard, unaltered wall rock and the brittle nature of ferro-axinite.

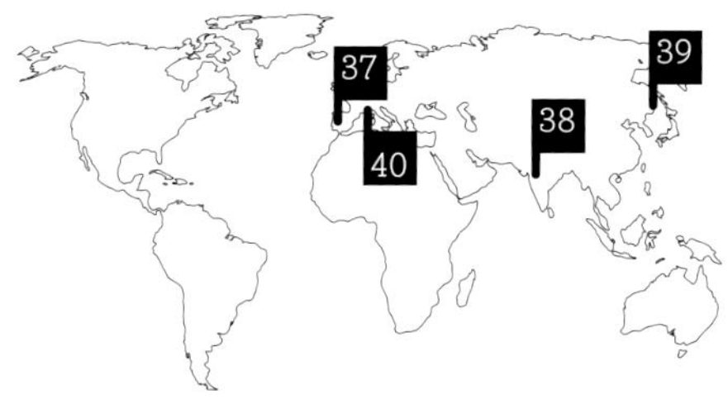

FLUORAPATITE

PANASQUEIRA MINE, PANASQUEIRA, CASTELO BRANCO DISTRICT, BEIRA BAIXA PROVINCE, PORTUGAL

The first tin-mining concessions at Panasqueira date from the late 1800s, but the veins there may have been worked sporadically as far back as Roman times. In 1928 the concession was taken over by Beralt Tin and Wolfram, Ltd., a British company (now controled by Avocet Mining), and after some years of development the mine began production in 1934; it is still in operation, and although past its peak, it is still considered one of the world's great tin-tungsten mines. The property actually encompasses hundreds of individual veins and a number of interconnected underground mines. These mines include the Panasqueira, Barroca Grande, Corga Seca, Vale da Ermida, and others; all feed out through the Barroca Grande and are collectively known today as the Beralt mine. The veins range up to a meter in width and frequently enclose open spaces where crystals of various minerals have been free to grow.

Panasqueira is world-famous for its specimens of arsenopyrite, wolframite, and fluorapatite, the latter in a range of color from very pale to dark green and from pale to dark purple. This specimen, comprised of a pair of large dark purple crystals, is exceptional.

FLUORAPOPHYLLITE

MOMIN AKHADA NEAR RAHURI, MAHARASHTRA, INDIA

Few regions in the world have produced such an enormous number of mineral specimens for the collector market as India has during the last three decades. Vast tracts of land covered hundreds of feet deep in cavity-rich basalt flows provide a virtually unlimited opportunity for new mineral discoveries, particularly for minerals of the zeolite group and related species. The many occurrences had long been noticed in passing, and at least one Indian collector, Jaime Ribiero (superintendant of railways in Bombay), was active in the 1940s. But Indian minerals drew no serious international interest until the 1960s and 1970s, when a few Indian mineral dealers began to make specimens available to the outside world. Since that time a number of Indian families have operated specimen businesses, supplying the world market. One of the earliest was Muhammad B. Makki, who began dealing in minerals in 1953.

In late 2002 a large circular well was being excavated in order to supply water for the village of Momin Akhada in Maharashtra State, about 50 km north of Ahmednagar. At a depth of 55 feet a large open cavity was intersected which proved to be lined with green fluorapophyllite, perhaps the most sought after of Indian minerals. The excavators, understanding the value of specimens, contacted Muhammad F. Makki (son of Muhammad B. Makki) and suggested that he come and have a look. The apophyllite was found growing in circular clusters on stilbite, an attractive and fascinating habit that had never before been observed anywhere. Many specimens were removed and were eagerly purchased by museums and private collectors, the example illustrated here being among the finest.

FLUORITE

NIKOLAEVSKIY MINE, DALNEGORSK, PRIMORSKIY KRAY, RUSSIA

The extensive mineral deposits around Dalnegorsk in far eastern Russia were originally discovered by the Chinese in 1872, when that area was under the control of Manchuria. In 1887 the Russians moved into the area, and ten years later began operations to extract copper, lead, and zinc. In 1910, following completion of a railway to the Sea of Japan, Dalnegorsk ore was shipped to the United States, Germany, and Holland. The British operated the mines for a few years, but in 1931 Russia nationalized the mines and Russian companies have continued to operate them ever since. The Nikolaevskiy mine, opened in 1982, is operated by the Dal'polymetal Company and is the largest ore producer of the eight major mines in the district.

The Nikolaevskiy mine has also produced some of the world's largest and finest colorless to pale green fluorite crystals. Some crystals with mirror-smooth faces are so perfectly clear and flawless inside that they almost seem to disappear. The pictured example is one of these, containing only a faint "phantom" inside representing an earlier growth phase.

LEFT: *Entrance to the Verchniy mine at Dalnegorsk.*

FLUORITE *on smoky quartz*

NORTH FACE OF LES DROITES, MONT BLANC RANGE NEAR CHAMONIX, HAUTE-SAVOIE, FRANCE

The French and Swiss Alps have been the source of beautiful minerals from countless small pockets, vugs, and clefts at least since the 1500s, when *strahlers* began to search for specimens as a part-time supplement to their regular occupations. The first legislation to govern mineral collecting was enacted there in 1759, and even before that the Swiss government would exact a portion of each important find as a tax. The tradition continues to this day, sometimes in families that have been climbing the high mountain cliffs for many generations.

Among the most sought-after and highly coveted of all Alpine minerals is pink octahedral fluorite, which has been found on both the Swiss and French

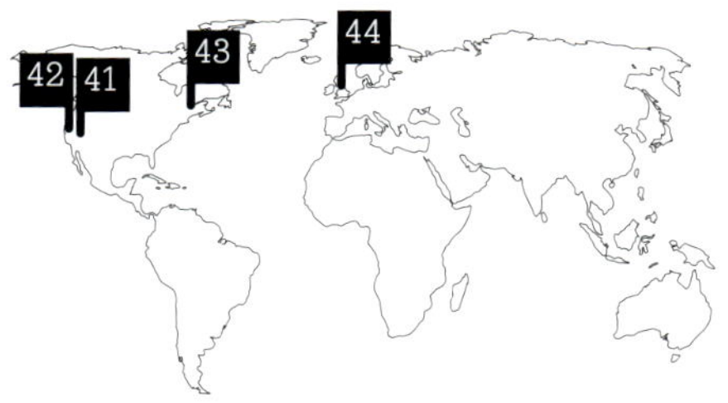

sides of the Alps. The crystals range from a very pale or grayish pink to bright and vivid pink to nearly red, and the crystals themselves range from sharp-edged and fresh to corroded in some degree – the most corroded being rather rounded and pitted and sometimes falling apart in pieces because of corrosion along cleavage planes. Pink crystals occurring in combination with smoky quartz, like the one illustrated here, are the most highly desired.

GOLD *named "The Dragon"*

COLORADO QUARTZ MINE, MARIPOSA COUNTY, CALIFORNIA

The Gold Rush of 1848–1849 made the Mother Lode area of California famous as a source not only of gold dust and gold nuggets but also crystallized gold. The early mining camp of Colorado, on Saxon Creek, sprang up following the Treaty of Guadalupe (1848), in which California was ceded to the United States by Mexico. Mexican placer miners who had worked the creek deposits for gold were driven out by the hoard of American forty-niners. Eventually the stream deposits were traced to their source veins, and in 1875 John Bataille filed a claim on the Colorado vein that came to be known as the Colorado Quartz mine. Bataille sold the mine to three men from the nearby town of Mariposa in 1885, and the property remained in the hands of their families until it was sold in 1974 to David Grimes, who sold it to Colorado Gold, Inc. of Fallbrook in 1979.

The Colorado vein, which varies from 1 to 7 feet in thickness, consists primarily of quartz and sporadically distributed pockets of crystalline gold. Long barren stretches are punctuated with exciting discoveries. In 1976, for example, a pocket containing 350 pounds of crystalline gold was found. The 18.5-cm specimen pictured here (which has been nicknamed "The Dragon" because of its shape) was found in 2000 by Bryan Lees.

GOLD

MOTHER LODE COUNTRY, CALIFORNIA

In January of 1848, James Marshall was supervising a work crew building a sawmill on the American River near Coloma for his employer, John Sutter. Looking down into the shallow water of the mill race he saw a glinting pebble of gold – and California would never be the same. Marshall's discovery sparked one of the largest human migrations in history, as half a million people from around the world descended upon California in hopes of finding wealth.

Shortly after Marshall's discovery Major Pearson Reading found gold in the Trinity River and General John Bidwell found gold in the Feather River. Soon the "Mother Lode" country was mapped out, an area in which gold was common in the rivers and streams, having washed down from weathering quartz veins in the higher elevations. Prospectors soon learned to follow the gold dust and nuggets upstream until it disappeared, knowing that within a short distance the hardrock vein of gold might be found. Inside those veins, undamaged by the effects of stream tumbling and abrasion, were beautiful, sparkling crystals of native gold, so attractive in its natural state that miners referred to it as "jewelry gold." The specimen pictured here is a prime example (plate 42).

GROSSULAR

JEFFREY MINE, ASBESTOS, QUEBEC, CANADA

Evan Williams discovered chrysotile asbestos on the Charles Webb farm 25 km southwest of Quebec City in 1881. Williams persuaded W. H. Jeffrey to open a mine on the site, paying Webb a royalty per ton of asbestos produced. A succession of operators worked the deposit in succeeding years, but it retained the name of its original operator and has always been known as the Jeffrey mine. Primitive mining and milling methods eventually gave way to increasingly sophisticated ones. As the open pit grew ever larger in size, locomotives were brought in to haul out the ore in 1920, and those were replaced by truck haulage in 1950. Since that time the pit has grown so large that part of the town of Asbestos has had to be relocated.

Fortunately the value of the beautiful grossular crystals and also the superb green to purple vesuvianite found there has long been recognized, and specimens have periodically been saved by miners and collectors when productive zones were blasted. Excellent specimens have been recovered during the last ten years, but the pictured example, with its two large and perfect crystals on matrix, remains one of the best ever found there.

LEFT: *The Jeffrey mine open pit, Asbestos, Quebec.*

HEMATITE

FLORENCE MINE, EGREMONT, WEST CUMBRIA, ENGLAND

The iron mines south of the town of Egremont in West Cumbria (formerly Cumberland) were opened following the formation of the Ullcoats Mining Company in 1897. By 1917 seven different shafts were operating along the vein, but some were short-lived because of intersected zones of sand that tended to flow into the workings like quicksand. The Millom and Askam Mining Company then took over and sank another shaft, ultimately taking out over 10,000 pounds of ore in 1902. An exploratory drilling campaign in 1914 revealed a huge deposit of hematite; the workings that were developed on this orebody were christened

the "Florence mine." Because of the disruptions occasioned by World War I, it was not until 1920 that the orebody was reached. Large dams were erected to keep water from flooding the lower levels. By 1940 the remaining ore reserves were low, so it was decided to highgrade some ore from around the shaft collar. A haulage drift was driven to the nearby Ullcoats No. 1 shaft, and the two mines were merged into the Florence and Ullcoats mine.

As a result of problems associated with nationalization, the mines were closed in 1968. In that same year the British Steel Corporation took over and cut drifts that joined up with the Beckermet mine. And thus the Florence mine grew to cover a length of 5 miles, but the enlarged operation was still unprofitable and so the company closed the mine complex in 1980. Botryoidal masses of black hematite known as "kidney ore" were common throughout the Florence mine. The specimen illustrated here is among the finest known examples (see plate 44).

HYDROXYL-HERDERITE

XANDA MINE, VIRGEM DA LAPA DISTRICT, MINAS GERAIS, BRAZIL

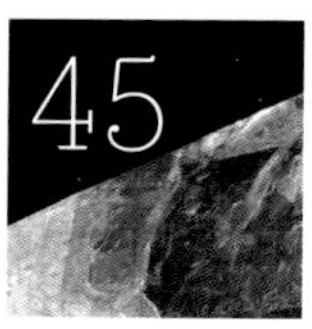

The Virgem da Lapa pegmatites in northeastern Minas Gerais, Brazil, were apparently first exploited for industrial-grade beryl in 1939. Production was sporadic over the years until 1974, when discoveries of attractive crystals of blue topaz, green tourmaline, and purple hydroxyl-herderite resulted in hundreds of free-lance miners swamping the area. For a short time during 1975–1976 the Xanda mine produced what have come to be regarded as the world's finest examples of hydroxyl-herderite. The Xanda crystals have an attractive grayish purple color in incandescent light, a bluish green color in daylight, and a tendency to occur in unusually large sizes. They are also very well formed, with sharp crystal edges and smooth, lustrous faces, and they tend to occur on a matrix of white feldspar crystals that makes for attractive specimens. Of the three mines exploiting the pegmatite bodies (the Xanda, the Limoeira, and the Toca de Onca), the Xanda was the most prolific producer, but by 1982 its various workings were either caved or filled with waste rock and inaccessible. No mining has taken place there since.

LEFT: *The Xanda mine, Virgem da Lapa, Minas Gerais, Brazil.*

ILVAITE

FIRST SOVIETSKIY MINE, DALNEGORSK, PRIMORSKIY KRAY, RUSSIA

There are eight mines in the area immediately surrounding the city of Dalnegorsk. The source of the great ilvaite specimens has been the First Sovietskiy mine, opened in 1934.

Dalnegorsk ilvaite is without question the finest in the world, and comes in a variety of habits, usually in association with colorless quartz crystals. The jet-black ilvaite crystals are razor-sharp and highly lustrous, usually with striated, low-angle, chisel-shaped terminations and with interestingly textured growth features on the crystal faces. Crystals up to 10 cm are known, but the 6-cm example shown here is exceptionally perfect and interesting in the way it stands up straight on the contrasting quartz matrix. (See entry 39 for history of Dalnegorsk region.)

JEREMEJEVITE

MILE 72 NEAR SWAKOPMUND, NAMIBIA

Colorless jeremejevite had been found as a handful of loose crystals at a Russian pegmatite in 1883. The first occurrence of blue crystals, however, was found a few hundred meters from the Namibian seacoast by Tannie Klippie Marais in 1973. The loose crystals she picked up eventually found their way to a local gem and mineral dealer, Sid Pieters, who recognized their potential value and had them positively identified. Pieters filed a claim covering the area of the discovery and brought in miners to excavate the granite in search of more crystals. The first specimens they found were colorless, but at the junction of two narrow pegmatite veins they finally found blue jeremejevite crystals associated with albite, quartz, lepidolite, apatite, and schorl. The largest pockets with the best blue crystals were all found within a meter of the intersection of the two veins. A second pocket of heavily altered blue crystals was found approximately 100 meters away in 1976.

The illustrated specimen was purchased from dealer Charles Key by Edward Swoboda in 1977. At the time, two of the crystals on the specimen were missing, but a search by Swoboda through the loose single crystals that Key had for sale turned up the missing two, and they were reattached to create the finest known specimen of the species.

LEFT: *Excavation for jeremejevite crystals at the Mile 72 locality, Namibia.*

LAZURITE

SAR-E-SANG MINE, KOKSCHA VALLEY, FIRGAMU, BADAKSHAN PROVINCE, AFGHANISTAN

Lapis lazuli is a mixture of minerals whose popularity extends back into ancient time; the deep blue lazurite reminded people of the night sky, the sprinkling of tiny pyrite crystals in the lazurite seemed like stars, and the white clumps of calcite marble resembled scudding clouds. The only significant locality for this material has always been the huge Sar-e-Sang mine in the remote Kokscha Valley of Afghanistan. Specimens dating to 3,000 BCE have been recovered from Egyptian archeological sites, and early Mesopotamian specimens are even older.

The well-formed crystals of lazurite found imbedded in marble, such as the example illustrated here, did not receive much notice as mineralogical specimens until recent times. In the 1980s specimens with the crystals partially cobbed out began to appear as Afghan miners and dealers discovered there was a market for them. Crystals to 1 or 2 cm are fairly common, and sharp dodecahedrons to 7 cm are known. Lately crystals of other minerals have been coming from the Sar-e-Sang mine, including crystals of pale blue afghanite and colorless crystals of sodalite that acquire rich purple color upon exposure to ulra-violet radiation (and sunlight).

LEADHILLITE

TSUMEB MINE, TSUMEB, NAMIBIA

Leadhillite is rare wherever it is found, including at Leadhills, Scotland, where it was first collected in the early 1800s. The occurrence found in the second oxide zone at the Tsumeb mine is unprecedented in the size, luster, and quality of the crystals. The 20 or so great specimens from that find, not all of which are particularly attractive, will probably stand forever as the finest examples of the species. (See entries 3 and 4 for history of Tsumeb mine.)

LEGRANDITE

OJUELA MINE NEAR MAPIMI, DURANGO, MEXICO

The Ojuela mine is well known to collectors worldwide as Mexico's greatest mineral locality. The complex orebodies were discovered by the Spanish in 1598 and have been mined almost continuously since that time. A total of 117 mineral species are known from the deposit, including the world's finest legrandite. The labyrinth of over 450 km of underground workings extending to a depth of over 900 meters is too dangerous and too confusing for any but longtime Ojuela miners to work there.

Legrandite was first found as yellow crystals to 4 cm at the Flor de Pena mine, Lampazos, Nuevo León, Mexico in 1932. Specimens were never common and very few reached the open market. The species reaches its finest development by far at the Ojuela mine near Mapimi, Durango, Mexico. The first significant specimens were found at Ojuela in the 1950s by American mineral dealer George Griffith, and again by him in 1961 and 1962. It has been found at numerous locations in the mine, from level 3 down to level 6, but the most spectacular specimens of all were found by Felix Esquivel in 1977, in the San Juan Poniente stope, level 5. The pocket measured roughly 30 x 45 cm and was filled with fabulous legrandite specimens that had already shaken loose from the walls. About 25 magnificent specimens, with slightly divergent crystal sprays up to 20 cm long, were recovered simply by reaching in and picking them up, including the 10.7-cm specimen illustrated here.

LUDLOCKITE

TSUMEB MINE, TSUMEB, NAMIBIA

The Tsumeb ore pipe, discovered in a remote region of German South-West Africa (now Namibia) in 1893, proved to be uniquely rich in rare elements such as germanium. In 1969 two American minerals dealers, Charles Locke Key and Frederick Ludlow Smith, found a boulder on the Tsumeb dump that proved to contain a mineral new to science. It was named after them in 1972, by combining their middle names to produce ludlockite. The boulder appeared to have come from the "germanite zone" just above the second oxide zone. Many years later a second occurrence in the third oxide zone on the 44 level yielded additional small specimens associated with white leiteite, reinerite, stolzite, and schneiderhöhnite. The illustrated specimen from the 1969 find is among the finest known examples of the species. (See entries 3 and 4 for history of Tsumeb mine.)

LEFT: *Tsumeb mine director Wilhelm Thometzek (in immaculate white suit) at the Tsumeb mine hoisting station, ca. 1920.*

MESOLITE *with fluorapophyllite*

PASHAN HILLS NEAR KHADAKVASLA, POONA, MAHARASHTRA, INDIA

Fine specimens of white mesolite in long, narrow crystals such as these to 20 cm or more have come from only one place in India: the Pashan Hills quarries near Khadakvasla, Poona, which are also the source of the world's finest green fluorapophyllite. Specimen mining under the careful direction of the Makki family has resulted in the preservation of specimens in perfect condition. Unfortunately the spread of residential areas near the quarries forced their closure in 1989 for

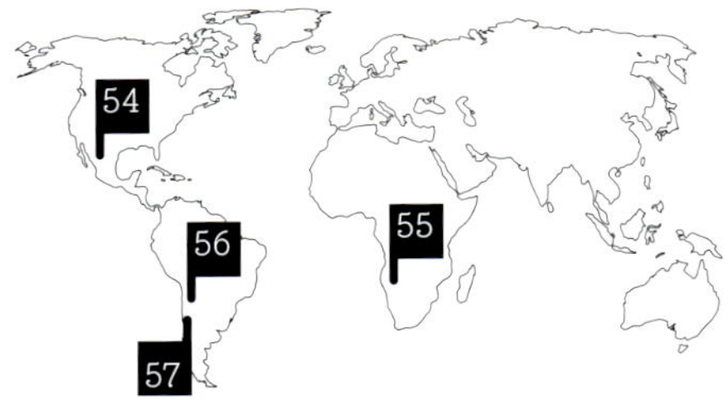

reasons of public safety; nevertheless some fine specimens were collected there after 1996 as well. (See entry 38 for more information on mining in India.)

MIMETITE

CONGRESO-LEON MINE, SAN PEDRO CORRALITOS, CHIHUAHUA, MEXICO

Mexico is littered with mysterious abandoned mines that beckon to the adventurous mineral collector willing to risk his life. In 1968 mineral dealer Benny Fenn decided to do some exploratory work in the Congreso-Leon mine, an extensive series of deep underground workings that have lain abandoned and slowly deteriorating since the time of the Mexican Revolution around 1912. The deepest workings reach 1,700 feet but the lower 600 feet of the mine were flooded. Furthermore, the timbering in the main shaft had been burned out at some time in the distant past, a hindrance that would have stopped most people. Fenn, however, lowered himself 900 feet on ropes and ladder fragments, getting off on a promising level where he ultimately found a large and beautiful pocket of bright yellow botryoidal mimetite. Being a skilled professional at mineral collecting, he extracted numerous fine specimens with very little or no damage. He wrapped and packed them carefully, then had to find a way to get them back up the 900-foot vertical shaft. Ultimately he had a hoist built to raise specimens so that he could continue to mine the occurrence until he had collected all that were to be found. Naturally the mineral world was excited to see exquisite specimens of beautiful mimetite, and the prices were such that even collectors of limited means could obtain at least a small example. Today the best pieces, like the 11.7-cm specimen shown here, grace major museum collections.

MIMETITE

TSUMEB MINE, TSUMEB, NAMIBIA

Mimetite has been found in a wide variety of spectacular and beautiful crystal forms and hues in the Tsumeb mine, and most of the different types were superior to mimetite from any other occurrence in the world until the recently discovered Chinese specimens joined the ranks of world-class quality. Fine specimens of any of the various Tsumeb crystal habits and colors would stand out in a connoisseur's collection, but the gemmy lemon-yellow crystals found in the second oxide zone in 1971 are still the ultimate for the species. (See entries 3 and 4 for history of Tsumeb mine.)

PHOSPHOPHYLLITE

UNIFICADA MINE, CERRO RICO DE POTOSÍ, POTOSÍ, BOLIVIA

Cerro Rico de Potosí is among the oldest and most famous orebodies in all of South America, and has yielded billions of dollars in silver. The Incas are said to have discovered it in 1462, and a Quechua Indian by the name of Diego Gualica rediscovered it for the Spanish in 1545. In the following centuries the Spanish extracted vast amounts of silver from dozens of rich veins that riddle the mountain, including enough to finance the 1588 Spanish Armada in its war with England. In the 19th century, as the exploitation of the near-exhausted silver reserves was declining, tin mining took its place, boosted by the high demand for tin caused by World Wars I and II. Production methods remained haphazard until several mines were unified with the Pailaviri to form the Unificada mine. In the 20th century, with the discovery of yet another rich silver vein, silver production resumed, along with the continuing recovery of tin, antimony, zinc, lead, and copper.

Despite the incredible silver output of Potosí, the mountain is famous among mineral collectors as the source of the world's finest specimens of a zinc-iron phosphate mineral, phosphophyllite. Specimens had been seen but not saved as early as the 1930s, but it was not until 1957 that a government chemist correctly identified the gemmy green crystals. The specimen pictured here, which is the second largest surviving crystal, was collected by Chief Geologist Aurelio Bustus in 1960. Relatively few other specimens have been preserved and all are considered extremely valuable.

LEFT: *Engraving of the silver-rich mountain known as Cerro Rico de Potosí, 18th century.*

PROUSTITE

DOLORES TERCERA MINE, CHAÑARCILLO, CHILE

Silver was first discovered in the Chañarcillo district in 1832 by a Chilean muleteer named Juan Godoy, who stumbled upon a vein of high-grade ore during a hunting trip. He became quite wealthy and soon other treasure seekers found rich outcrops in the same area. By 1850, nearly 1,750 miners were working 18 productive mines in the district. Production using hand labor amounted to over 10 million pounds of ore per year. The Dolores Tercera mine was especially productive of beautiful specimens of proustite in crystals exceeding 2 cm. Today the old workings are long abandoned, and nothing remains of the once bustling mining town of Juan Godoy.

Although the richness and crystallinity of the silver ores at Chañarcillo apparently rivaled those of the classic German silver deposits, relatively few specimens of silver minerals were preserved. Those that still survive from the early days of mining are among the most coveted of all specimens in the mineral world today, including the illustrated example (see plate 57).

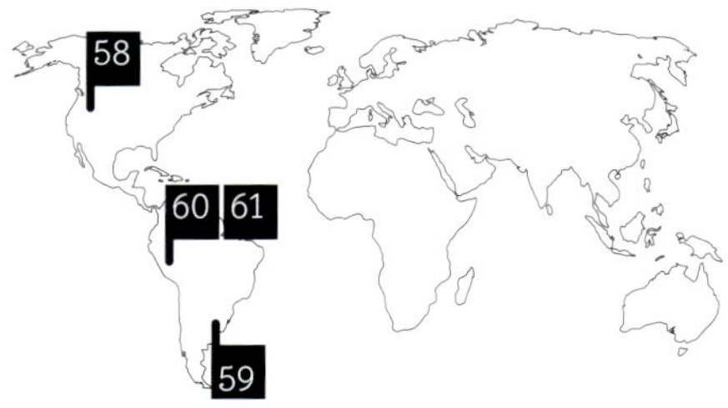

PYROMORPHITE

BUNKER HILL MINE, KELLOGG, COEUR D'ALENE REGION, SHOSHONE COUNTY, IDAHO

Thomas Irwin first staked a gold claim in the Coeur d'Alene region in 1878, and by 1884 a number of lead-silver veins had been found. The Bunker Hill vein was discovered in 1885 by Noah Kellogg. Eventually the Bunker Hill mine was expanded to absorb other mines including the Sullivan, Arizona, Bluebird, Stemwinder, Tyler, and many others. Mining has been under way continuously at Bunker Hill since 1887 – a remarkable record in itself.

In the early years of mining at Bunker Hill, many fabulous specimens of cerussite and some of pyromorphite were temporarily saved by miners and mine managers, but eventually nearly all were discarded before their mineralogical significance became known. Fortunately a major discovery of pyromorphite took place in 1980 in the Deadwood-Jersey vein of the Bunker Hill mine. A series of connected pockets yielded thousands of superb pyromorphite crystal clusters in colors of green, yellow-green, and orange. The 11.6-cm specimen pictured here is from that find and represents the best pyromorphite specimens ever found in North America.

LEFT: *Entrance to the Kellogg tunnel, Bunker Hill mine.*

QUARTZ *variety amethyst*

ARTIGAS, ARTIGAS DEPARTMENT, URUGUAY

The amethyst deposits of Uruguay occur in a southern extension of the basaltic rocks and lava flows that cover vast areas of Rio Grande do Sul, in southern Brazil, and are the source of most large amethyst geodes seen on the market. In very rare cases large calcite crystals have been found partially coated by small amethyst crystals, with the amethyst preferentially covering certain faces in an interesting design. Amethyst is one of the most widely available minerals on the world market, but perfect, symmetrical stalactites of amethyst like the one pictured here are exceptionally rare.

QUARTZ *Scepter*

BERILO BLANCO MINE NEAR SAPUCAIA DO NORTE, MINAS GERAIS, BRAZIL

The Berilo Blanco ("White Beryl") mine, owned by Sebastião Ferreira, was the world's first source of beautifully crystallized pink quartz. In early 1959 his miners encountered the first small pockets and stringers with smoky quartz crystals, then pockets containing large muscovite sheets and stocky quartz prisms. Behind these pockets the main pink quartz pocket was found, and it proved to be spectacular. The vug measured about 6 inches across, about a foot tall, and extended for over 16 feet. It was lined with large clusters, fans and sprays of exquisitely beautiful pink quartz crystals, some of them perched on or wrapped around large, pale smoky quartz crystals. Over 200 specimens including at least 50 major pieces were recovered and many of them were sold to California collector/dealer Peter Bancroft. The mine closed shortly thereafter, around 1961, but for the next 30 years continued to yield occasional small lots of specimens to part-time diggers, including the 25-cm scepter crystal pictured here.

QUARTZ *with siderite*

MORRO VELHO MINE, NOVA LIMA COUNTY, MINAS GERAIS, BRAZIL

Mining began at the Morro Velho gold deposit in 1834, first as an open pit. Production nearly ceased following a catastrophic fire in 1867, but in the 1870s mining resumed as an underground operation following the plunging orebody. The mine now reaches a depth of more than 3,000 meters (9,000 feet), and for a time was the deepest mine in the world. The rock temperature in the lower levels exceeds 130° F, making it impractical to mine any deeper.

The Morro Velho mine began producing significant numbers of mineral specimens in the 1960s, and until 1970 the mine management allowed miners to collect and bring out specimens to sell. As the specimens came under increasing demand, more and more collectors and dealers came into Nova Lima looking to buy specimens. Accidents occurred in the mine because miners became overenthusiastic about suspending operations in order to collect whenever pockets were encountered. Consequently, in 1970 the mine management prohibited collecting. The only result of this was that miners collected smaller specimens, which were easier to carry out unseen.

As this specimen illustrates, the mine is famous for attractive specimens composed of gemmy brown siderite, bronzy pyrrhotite, and beautiful quartz crystals (see plate 61).

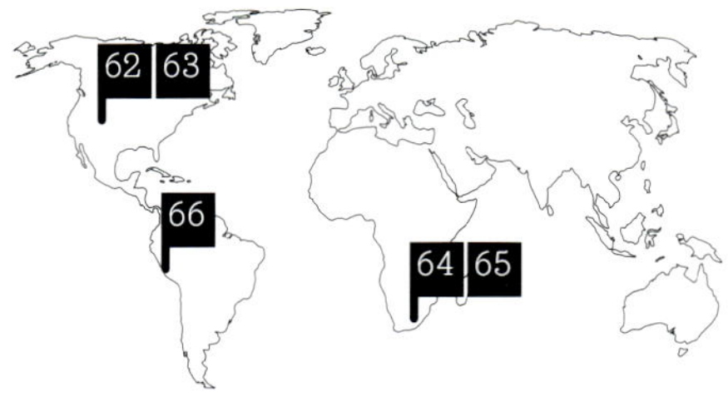

RHODOCHROSITE *named "The Alma Queen"*

SWEET HOME MINE NEAR ALMA, COLORADO

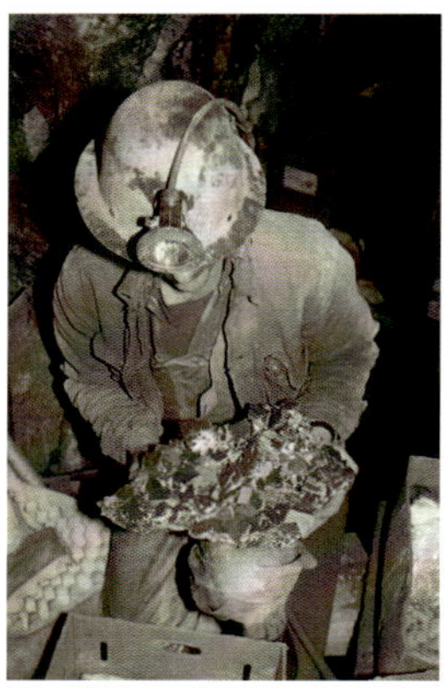

The Sweet Home vein complex was discovered in 1873 and the claims patented in 1876, among the earliest patents granted under the General Mining Law of 1872. Situated at an elevation well above 10,000 feet, the Sweet Home mine was opened as a silver mine. Over the years numerous gold and silver claims have been staked throughout the area, and the region has enjoyed a rich history featuring the colorful characters, lively boom towns, and turbulent existence common to most early mining districts.

This rhodochrosite was acquired from the miners by Peter Bancroft in 1966 (for about $6,500) and became widely known as "The Bancroft rhodochrosite." David Wilber purchased the specimen in 1974 and sold it to Ed Swoboda in 1979 (in a cash/trade deal that involved a Rolls Royce). Once pictured in *National Geographic* magazine, this famous specimen was valued at $85,000 in 1984 when it was acquired by the Houston Museum of Natural Science as part of the Perkins Sams collection. It is worth many times that amount today.

LEFT: *Large rhodochrosite crystal cluster just recovered from "The Good Luck" pocket in the Sweet Home mine, 1992.*

RHODOCHROSITE

SWEET HOME MINE NEAR ALMA, COLORADO

The Sweet Home mine was a minor producer of silver, but its true claim to fame rests upon the exceptional rhodochrosite crystals found there. Fine rhodochrosite specimens were noted as early as 1876, and specimens were being collected and sold from the 1870s through the early 1890s. A major find was made in 1925, from a crystal-lined cavity 30 feet long, and specimens from it made their way to collections throughout the world, including the American Museum of Natural History's J. P. Morgan exhibit in Paris. The mine has been active for the last 13 years, specifically for the production of fine mineral specimens, under the direction of Bryan Lees of Golden, Colorado, who recovered this piece.

RHODOCHROSITE *on manganite*

N'CHWANING MINE, KURUMAN, NORTHERN CAPE PROVINCE, SOUTH AFRICA

The Kalahari Manganese Field in northern Cape Province, South Africa, is among the world's great ore deposits. Hidden beneath the sands of the Kalahari Desert, which spans hundreds of miles across north-central South Africa, are the world's largest known deposits of manganese. In the 1920s prospectors found manganese under the sands surrounding a small ridge known as "Black Rock," the only surface evidence of the massive ore body below. The Black Rock area was first described by L. G. Boardman in 1941, and between 1951 and 1954 the enormous 56-km manganese orebody was prospected by geophysical methods and drilling. Soon thereafter mining commenced where the sand cover was thinnest. The N'Chwaning mine was opened in 1974 and proved to be especially rich, not only in manganese ore but also in world class specimens of rhodochrosite, found shortly after the mine opened, at a depth of 30 to 40 meters. Specimens of major importance began appearing in the international mineral market in 1977. Associated Manganese Mines of South Africa Ltd. owns the N'Chwaning mine, directed fortunately by an avid mineral collector, Desmond Sacco, who has encouraged the preservation of specimens.

Rhodochrosites from the N'Chwaning mine come in a variety of habits, from gemmy red scalenohedrons (plate 64) to stubby clusters, gemmy spheres covered with crystal points, flattened translucent crystals, zoned barrel-shaped aggregate crystals, and frosty spherical formations of a pale pink color (plate 65), often associated with black manganese oxide minerals.

RHODOCHROSITE

UCHUCCHACUA MINE, OYON PROVINCE, LIMA DEPARTMENT, PERU

The Uchucchacua mine is located at an elevation of over 15,000 feet in the Andes Mountains of west-central Peru. Mining there dates from Spanish colonial times in the 16th and 17th centuries, as evidenced by the ancient workings, all of which are in the upper, oxidized zone. However, no written records appear to have survived from those early days. Modern mining appears to date from 1975 when a pilot plant for processing 150 tons of ore per day was installed. In 1981 three large, high-grade orebodies were discovered, and production was increased to 500 tons per day with an average silver content of 16 ounces per ton. Uchucchacua became the third largest silver producer in Peru, with an output of 3 million ounces per year.

The ore consists of veins and massive orebodies that formed in faults and adjacent wall rocks. Carbonates and sulfides occur along reaction fronts replacing manganese silicates, resulting in beautiful specimens of the manganese

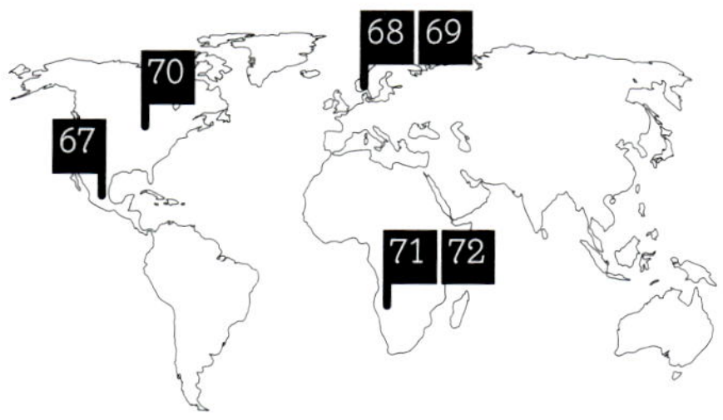

carbonate, rhodochrosite, in crystals up to 4.5 cm such as the example seen here. The dark, splotchy, manganese-rich matrix is characteristic for the locality.

SCORODITE

EL COBRE MINE, ARANZAZU, MUNICIPIO DE CONCEPCIÓN DEL ORO, ZACATECAS, MEXICO

Long before the Spanish discovered the enormous ore deposits at Zacatecas in 1568, the local Indians were collecting gold and silver from outcrops in the high mountains. A few miles away another major mining camp, Concepción del Oro, grew up on the western edge of the Zacatecas Plateau. Just beyond the town are the Temeroso Mountains, in the high reaches of which lie the small, now-abandoned mining villages of El Cobre and Aranzazu. During the mid-1890s illiterate local contractors took over the mines there, leaving no written records. Nevertheless, some fine specimens of azurite and malachite came out of the El Cobre mine in those days and found their way onto the specimen market. The locality would have remained relatively obscure but for a single unique pocket of crystals found by a miner at the El Cobre mine in 1968.

He removed the specimens and sold them to a wholesaler in El Paso, Texas, who sold all of the pieces very quickly and very quietly. Today the specimens reside in the major museums of the world as well as in the collections of a few lucky private collectors. The beautiful 6-cm specimen shown here is among the finest from that find.

SILVER *with calcite*

KONGSBERG, NORWAY

The Kongsberg silver deposits in southern Norway were discovered by two young herders in July of 1623. Their families tried to keep the find a secret, but when they were arrested for trying to sell presumably stolen silver they had to go public with their discovery. Within two years there were four mines operating in the area, including the Kongens Grube ("King's mine"), which proved to be the most productive in the district and eventually reached a depth of 1,076 meters. The nearby mining town that sprang up was named Konningsberg ("King's Mountain"), later shortened to Kongsberg. Mining continued for over 330 years, finally closing down in 1957 after producing over 43 million troy ounces of silver.

The Kongsberg mines have always been the world's premier source of silver specimens, as coarse wires and also as crystals. The various wire silver habits have been called arborescent silver, tooth silver, ram's horn silver, hair silver, moss silver, and thread silver. Coarse, unstriated wire silver bound by obvious tiny crystal faces has been called baroque silver. The two specimens pictured here are superb examples of heavy wire silver with associated calcite and sphalerite.

LEFT: *A miner cobbing native silver ore in the Kongsberg mine, ca. 1900.*

SILVER

QUINCY MINE, PEWABIC LODE, HOUGHTON COUNTY, MICHIGAN

Native copper and probably also native silver on the Upper Peninsula of Michigan were well known among the local Indians for thousands of years before the occurrences first came to the notice of Europeans. Since the first claims were staked in 1844, the mines have been considered copper mines, and for a long time the silver that was seen periodically was considered by miners to be theirs for the taking, as a kind of bonus. Thus many fine silver specimens were preserved by miners and their families over the years. At least a small amount of silver has been found in every copper lode and fissure in the district. Production from the peninsula had peaked by the time of World War I, and in the following years the mines gradually closed, one by one. Over 13 billion pounds of copper and 16 million ounces of silver were produced since 1845.

The Quincy mine, source of the silver specimen pictured here, opened around 1856 following the discovery of the Pewabic lode by C. C. Douglass at an outcrop that had been worked by ancient Indians. Mines on this lode proved to be highly productive, and the last one did not close down until 1967. A half million dollars in silver is officially recorded as having come from the mine, but probably much more was stolen and not recorded, accounting for the fairly large number of specimens that have been preserved.

LEFT: *The Number Two shaft-rockhouse at the Quincy mine, ca. 1895.*

SMITHSONITE

TSUMEB MINE, TSUMEB, NAMIBIA

Smithsonite was abundant in the oxidized zones at Tsumeb, in an extravagant variety of colors. Old specimens from the first oxidation zone are generally the least attractive (brown, yellow, and white crystals), whereas the crystals like these found in the second and third oxidized zones are more spectacular, in shades of green (plate 71), pink, turquoise-blue (plate 72), and colorless. (See entries 3 and 4 for history of Tsumeb mine.)

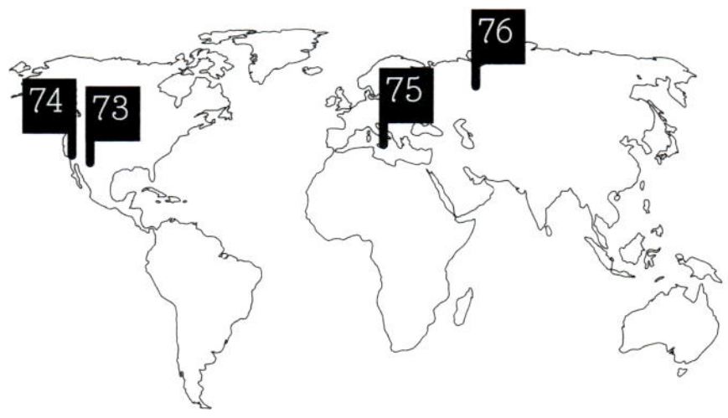

SPANGOLITE

CZAR SHAFT, COPPER QUEEN MINE, BISBEE, COCHISE COUNTY, ARIZONA

By the mid-1970s, when the Bisbee mines shut down, the district had produced nearly eight billion pounds of copper and more lead, silver, and gold than any other mining camp in Arizona. For over 100 years, this mining district continually produced magnificent specimens of azurite, malachite, cuprite, and more than 200 other species in a range of habits, including this rare specimen pictured from the Czar shaft.

The Czar shaft, opened in 1885, was the first to be sunk by the Copper Queen Mining Company. More fine mineral specimens (mostly carbonate minerals such as azurite and malachite) came from it than from any other mine in the district. It was closed in 1944. (See entry 5 for history of Bisbee mines.)

LEFT: *A mine mule hauling ore cars in the Czar mine at Bisbee, 1908.*

SPODUMENE *variety kunzite*

PALA CHIEF MINE, SAN DIEGO COUNTY, CALIFORNIA

The Pala Chief mine is a patented claim situated on the northwest side of Chief Mountain in the Pala District of San Diego County, California. It was opened in 1903 by Frank Salmons and others, and for the next ten years was the world's leading producer of gem-grade kunzite (lilac-colored spodumene). Although kunzite was first discovered on neighboring Hiriart Mountain, the crystals from the Pala Chief mine are the finest in the world. It is interesting to note that Pala Chief kunzite is generally twinned, whereas kunzites from the other nearby localities such as the White Queen mine are usually not, and this factor can be used to help distinguish specimens from the localities.

The Pala Chief mine is currently owned by Bob Dawson, who has improved road access, driven additional workings, and found excellent specimens of kunzite to 11 cm and pink elbaite (a tourmaline). The 30-cm crystal shown here is an historic specimen found shortly after the turn of the century by Frank Salmons.

LEFT: *Bob Dawson, owner of the Pala Chief mine, examining part of the pegmatite vein.*

SULFUR

COZZODISI MINE, AGRIGENTO PROVINCE, SICILY, ITALY

Many sulfur mines operated in Sicily in the 19th century. The Cozzodisi mine was opened in 1870 as a surface-mining operation, and underground workings were initiated in 1882. Mining reached its peak in the late 1800s because of the high demand for sulfur for the production of industrial sulfuric acid. Catastrophic fires, such as the one of 1887, which was not completely extinguished until 1891, regularly interrupted mining. In 1911 a third underground level was opened as

the upper levels became exhausted or burned out, and eventually 12 levels to a depth of 300 meters were opened. The worst mine fire began in July of 1916, and claimed the lives of 89 miners. Mining operations were suspended in 1988 because the mine could not be adapted to the American Frasch process of sulfur mining, untilizing superheated water to melt the sulfur. Sulfur mining in Sicily was always carried out by hand, using relatively primitive methods that allowed maximum opportunity for collecting of specimens such as this along the way. Because of these inefficient mining methods, Sicily's mines could not compete with the American salt dome deposits. Today all of the Sicilian mines are closed.

LEFT: *Abandoned surface buildings at the sulfur mines in Sicily, ca. 1909.*

TOPAZ

ALABASHKA, ALABASHKA-MURSINKA PEGMATITE FIELD,
YEKATERINBURG OBLAST, RUSSIA

The Alabashka-Mursinka pegmatite field is the most famous granitic pegmatite region in all of Russia and the former Soviet Union, being most noted for superb gem crystals of beryl, topaz, and elbaite (a tourmaline). The pegmatite fields around the village of Alabashka in the Ural Mountains were the classic source of blue topaz – although yellow and colorless topaz occurs there as well – for over two centuries, until the discoveries in Brazil were made. The earliest known record of the gem mineral occurrences in this area is a 17th-century report made to the Siberian Ministry in Moscow by Dmitri Tumashev. He performed the work at his own expense, then received a mere pittance in compensation after delivering the report in 1668. Gem mining, however, was not really developed to a significant scale until the mid-18th-century; state-sanctioned operations began in 1775.

Blue topaz is one of the great classics from Russia, especially considering that no new specimens of note have been found in many decades. Most great specimens, like the one pictured, are now found only in major museums.

LEFT: *Excavations into a pegmatite near Mursinka, in the Ural Mountains, 1915.*

TOPAZ

ANTÔNIO PEREIRA MINE, ANTÔNIO PEREIRA, DOM BOSCO, MINAS GERAIS, BRAZIL

For over 200 years the area around Ouro Preto, Brazil, has been a source of mineral riches, including gold, iron, manganese, and "Imperial" topaz of a beautiful reddish golden-orange color, such as this example. Topaz, originally called "Brazilian ruby," was discovered there in 1751, from a series of deposits that run east-west for about 25 km. The find caused great excitement in the royal court in Lisbon (Brazil was at that time a Portuguese territory), resulting in the stones being referred to as "Imperial." A number of the deposits were worked intensively for many years, some to exhaustion. In 1908 a new topaz mine was opened

at Saramenha (also called Vermelhão), a locality that has proven to be particularly rich in large, well-formed crystals beloved by connoisseurs.

Since then work has been sporadic, with most mining in the district carried out by traditional, primitive hand labor, except at the Saramenha and Capão mines, which have been successfully mined with mechanized equipment for several years. The easternmost mine in the district, and the source of the illustrated specimen, is the Antônio Pereira mine.

TORBERNITE

MUSONOI MINE NEAR KOLWEZI, KATANGA, DEMOCRATIC REPUBLIC OF THE CONGO

Uranium was first discovered in the Katanga Province in 1913, at the Luiswishi copper mine, and subsequently also at a number of other cobalt-copper deposits. Uranium minerals were found at the Shinkolobwe copper mine in 1915, at the Kalongwe mine in 1930, and at the Musonoi mine in the 1950s. Only at Shinkolobwe were the uranium minerals plentiful enough to be mined as ore (Shinkolobwe uranium was used in the Hiroshima atomic bomb). At the other localities the amounts present were of little economic importance, although they spurred great mineralogical interest.

In the western sector of the Musonoi copper-cobalt mine, a zone measuring 50 x 100 meters was found to contain an abundance of copper-uranium minerals, including the world's finest examples of torbernite and also another uranium-copper mineral, cuproslodowskite. The torbernite occurs as large, dense clusters of lustrous, flat, square crystals up to 3 cm each. The example shown here is typical of the finest specimens (see plate 78).

ZOISITE *variety tanzanite*

MERELANI HILLS NEAR ARUSHA, UMBA VALLEY, TANZANIA

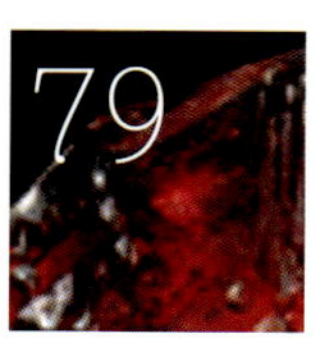

Originally recovered from the Merelani Hills during mining operations in Tanzania, this exceptional tanzanite crystal is one of the largest and finest gem quality crystals ever found. Because the crystal is exceptionally *clean* (lacking internal flaws) it is even more remarkable that this crystal was not cut up into gemstones for use in jewelry. This crystal is important not only for its gem value, but also because it is an exceptional example of *trichroism* ("three colors").

When viewed from the top (looking down the c-axis) the crystal exhibits a deep plum red color; it appears to be sapphire blue when viewed from one side (perpendicular to the c-axis); and it displays a rich purple color when rotated 90 degrees around the c-axis. Trichroism in tanzanite crystals also provides evidence that the crystal has not been heat-treated. Gemstone miners, dealers, and cutters routinely heat tanzanite crystals to make them appear more blue, satisfying the demand of the international jewelry market; doing so destroys the natural trichroism of the crystals.

Discovered in 1991, this crystal changed hands several times belonging to various gem dealers and collectors in Tanzania, Thailand, Germany, and the United States before it was acquired by the Houston Museum of Natural Science in the late 1990s.

ZOISITE *variety tanzanite*

MERELANI HILLS, SIMANJIRO DISTRICT, LELATEMA MOUNTAINS, ARUSHA PROVINCE, TANZANIA

In July 1967 a tailor named Manuel d'Souza from the Indian province of Goa was prospecting for rubies in Tanzania. He enlisted the help of local Masai tribesmen, who were unaware of any deposits of red stones, but who led him to an outcrop where gemmy blue crystals could be found. D'Souza initially thought the crystals were sapphires, but laboratory analysis revealed the blue crystals to be a hitherto unknown gem variety of the mineral zoisite. He registered four claims with the mining office, and shortly thereafter a Greek Army officer by the name of Papanicholau filed more claims, and soon the site on Merelani hill was riddled with new prospects and diggings. Tiffany's in New York became involved in promoting tanzanite on the international gem market. Conflicts and court actions followed, until the mines were nationalized in 1971 under the name Tanzanian Gemstone Industries, which was taken over by the Tanzanian State Mining Corporation in 1976. As often happens when mines are nationalized in third world countries, efficiency and production fell. For a time in the late 1980s the government lost control of the area and thousands of illegal miners flocked in, but in 1991 control was regained, and a licensing system involving four concessions along the vein (known as Blocks A, B, C, and D, Block C being the largest producer) totaling about 5 square miles has been established, which discourages large-scale exploitation by foreign mining companies. In 2000 the African Gem Resources Corporation (AFGEM) was given the rights to mine two-thirds of the known tanzanite reserves. Working conditions are hard, however, and the mines are closed for several months every year because of flooding. Nevertheless, many new stones are mined each year, and tanzanite has become the most popular faceted stone after diamonds, emeralds, rubies, and sapphires. Very large crystals such as the specimen illustrated here are rarely saved from the cutter (see plate 80).

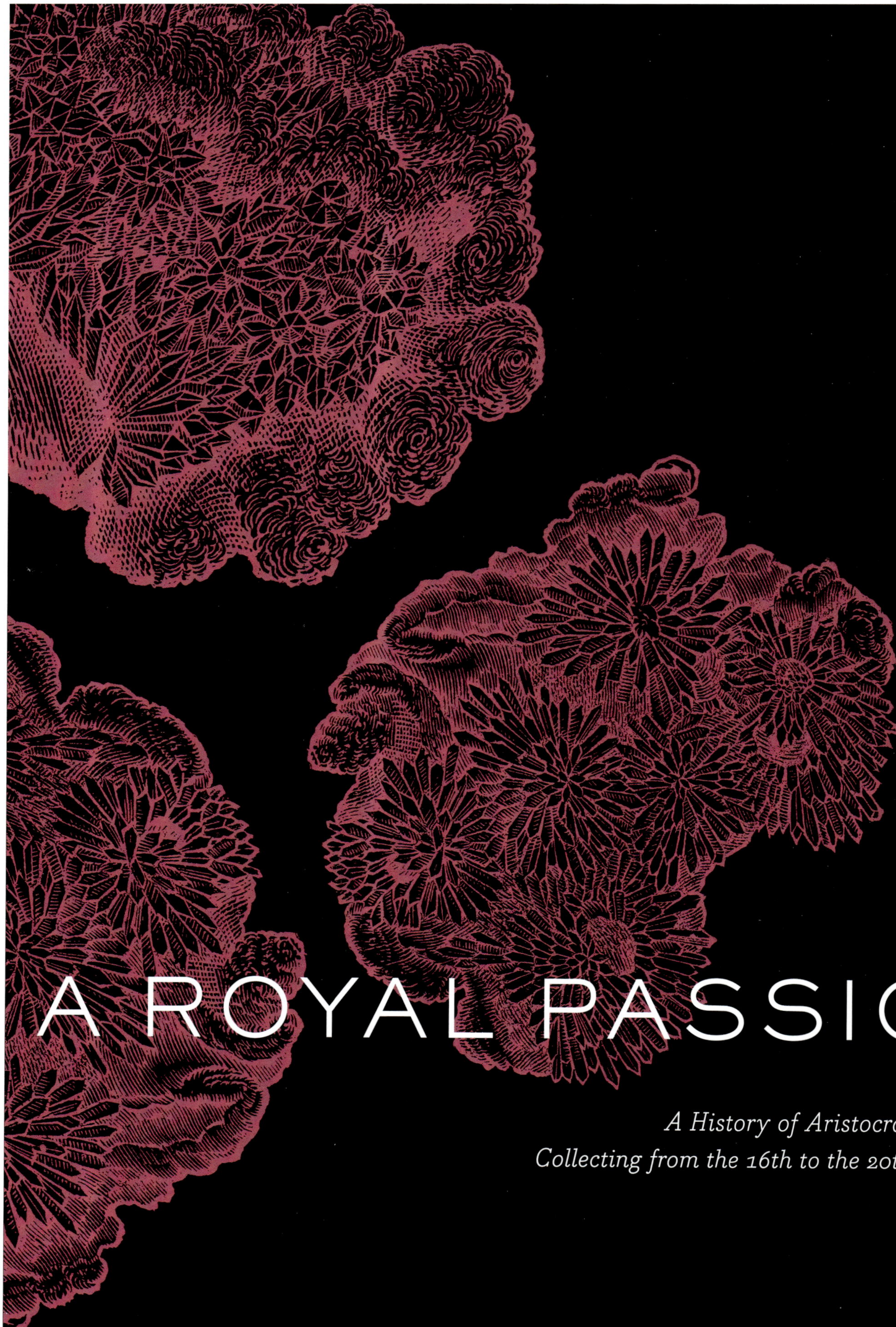

A ROYAL PASSION

A History of Aristocratic Mineral Collecting from the 16th to the 20th Centuries

If horse racing is "the Sport of Kings," then mineral collecting must surely be "the Hobby of Kings" – and of emperors, empresses, dukes and archdukes, princes, counts, and barons as well. At least that is what history would suggest. And yet mineral collecting is also the most egalitarian of hobbies. A person with no money at all can go out into the field and find free specimens, pulling them out of the ground, while people of greater means can do the same or buy specimens. Royalty naturally preferred the best, regardless of cost, and over the centuries built some incredible collections that are today national treasures in public museums.

Over the years and the centuries, of course, the prominence in society of the aristocratic class has dwindled. But they left a legacy of collections and of scholarly interest and philanthropic support that continues to benefit and inspire society. Today, in the 21st century, it is primarily professional people of the business class who build the great collections and support the great museums, along with their public-spirited corporations, following a tradition taken from their aristocratic predecessors. Nevertheless, building great mineral collections today continues the same scholarly and ultimately philanthropic tradition that has existed since the days of Michaelangelo and DaVinci.

The RENAISSANCE

Humans have probably been picking up and saving interesting bits of rock and mineral material since the Stone Age. However, it was the Renaissance that profoundly awakened public interest in Europe, and a serious desire arose to learn the truths about nature. The driving force for this resurgence of interest in the physical world was the rise to prominence of the scholar as a third leading power in society (along with the clergy and the aristocracy). Scholars, according to the collector and physician Johann Major (1636–1683), achieved their fame by collecting alone, and not by birth, so they needed a proper environment to carry out their studies; thus was born the modern notion of a museum. Scholars, however, often needed the support of an aristocratic patron to carry out their work and to assemble proper collections; thus was forged a partnership that inspired and enriched the aristocracy as well, and ultimately benefited all members of society.

How should such a museum be furnished? Inspiration was provided by Pliny in his ancient *Historia naturalis*, written around 79 AD and preserved as carefully copied, recopied, and translated manuscripts. In 37 volumes Pliny describes the entire natural world as far as it was known in his time, including much commentary on minerals. According to Pliny's philosophy, the goal of all human research into natural history is to realize fully the divine causality of all existence, and the humble triviality of man in comparison to the infinite power of nature. To Pliny, the "recording, analysis and investigation [of nature] represented the highest goal of all human activities" (Schulz, 1990). To be properly studied, specimens of the constituents of nature had to be assembled in one place, and Pliny's work was taken as a guide for the building of such collections.

Although initially conceived as primarily a theological exercise, the study of natural history soon began evolving along progressively more scientific lines. Scholars began to question the purported supernatural properties of minerals, and to show more interest in their true physical characteristics.

The change was evident in the 16th century's first mineralogical work, *Speculum Lapidum* ("The Mirror of Stones"), published in 1502 by Camillus Leonardus. Here, finally, an author begins to consider the physical properties of minerals (transparency, hardness, density, porosity, etc.) that can be used in the identification and characterization of species. With Leonardus there is therefore some reasonable basis to speculate that he may have had a study collection of minerals, although there are no written statements to confirm it.

Leonardus was court physician to Cesare Borgia (1475–1507) in the seaport city of Pesaro, in central Italy. Because of the necessity for physicians to be educated in the practical aspects of nature, including the pharmaceutical uses of minerals and plants, the combination of physician (or apothecary) and naturalist became common and persisted into the 19th century. Aristocrats often supported the building and curating of their own pharmaceutical collections (should His Highness become ill, it was wise to have a wide range of pharmaceutical materials at hand); the partnership between scholar and aristocrat thus often began for such practical reasons. It was just a matter of time before pharmaceutical minerals and metallic ores began to excite curiosity and interest for their own sakes in the minds of Renaissance scholars and their aristocratic patrons.

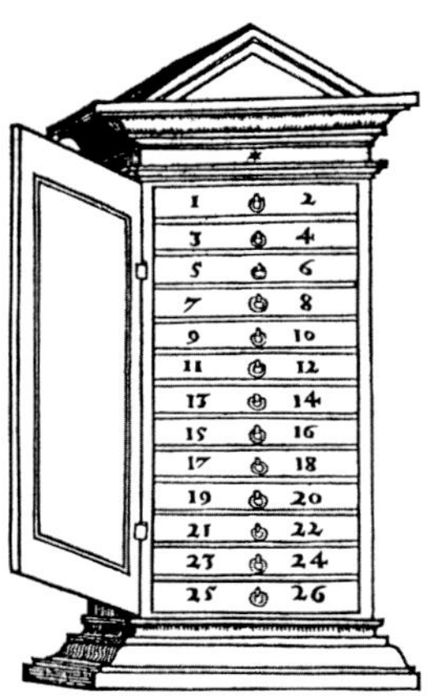

Mineral Cabinet of Johannes Kentmann

The 16TH CENTURY

Mineral collecting began in the 16th century, in a form strikingly mature and similar in concept to modern collecting. That is, collectors such as Georgius Agricola, Johannes Mathesius, Johannes Kentmann, and Bernard Palissy *began* as specialists in mineralogy and not as general natural history collectors. Most of their successors for at least the next two centuries tended to collect botanical and zoological samples as well, rather than specializing so intensely in mineral products alone.

Collections of the nobility were decidedly general instead of specialized. They grew by the accumulation of every conceivable type of object, natural and artificial, without any scientific purpose but more to illustrate or symbolize man's central place in the universe, and the sovereign's dominion over all.

Scholarly collections of a general nature included minerals with plants and animals, reflecting a more intellectual interest in nature. Conrad Gesner, the prominent Zürich naturalist, spent most of his life collecting and writing about plants and animals, turning to mineralogy only in his last years. Private Italian collections of *naturalia*, developing in parallel with the princely collections, also served a more didactic purpose. Their owners, lacking high aristocratic backgrounds, studied medicine and botany at universities and went on to professions as pharmacists (e.g., Francesco Calzolari and Ferrante Imperato), physicians, professors, and directors of important botanical gardens (such as Michele Mercati and Ulissi Aldrovandi). They, like Agricola, began studying minerals as *materia medica* but developed a more consuming interest and a compulsion to collect all manner of mineral specimens regardless of their potential pharmacological value. For these scholars, building a personal collection of plants, animals, and minerals meant acquiring an invaluable store of research specimens for first-hand laboratory study. Furthermore, the open access that students and the public had to private museums such as those of Aldrovandi and Calzolari sets them firmly apart intellectually from the princely collections, which were guarded, private preserves. They differed even in their arrangement; the scientific museums were organized rationally like file cabinets, and the princely collections (with the exception of that of Augustus of Saxony) followed arcane and symbolic arrangements of often mysterious significance (Olmi, 1985).

Elector Augustus of Saxony

By the latter half of the 16th century, a whole range of *naturalia* had found its way into private collections, especially in Italy. Encyclopedias of natural history began to appear as guidebooks for collection builders. Samuel Quiccheberg (1529–1569), personal physician to Duke Albrecht V of Bavaria, produced a particularly thorough treatise, considered also to be the earliest known museological tract (*Inscriptiones vel Tituli Theatri Amplissimi . . .*, published in Munich in 1565). He carefully categorized the contents and organization of an ideal museum and library, covering all types of artwork, plants, animals and minerals – a "rich theater of objects of the whole universe." He employed, for the first time, the term *Wunderkammer* (repository of extraordinary objects), and he concurred with Pliny in emphasizing the great importance of preparing collection *catalogs*.

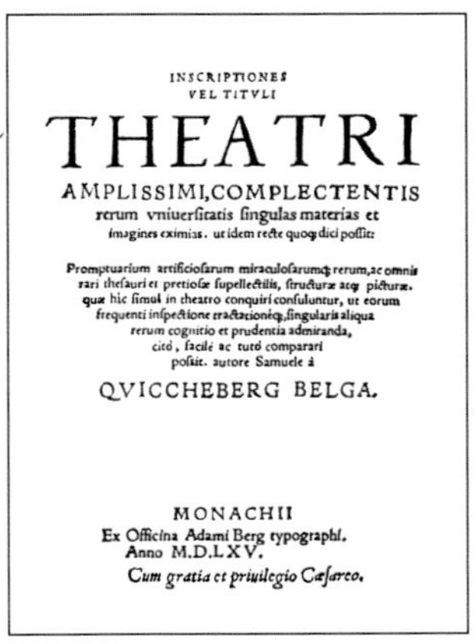

INSCRIPTIONES
VEL TITVLI
THEATRI
AMPLISSIMI, COMPLECTENTIS
rerum vniuersitatis singulas materias et
imagines eximias. ut idem recte quoq; dici possit:

Promptuarium artificiosarum miraculosarumq; rerum, ac omnis
rari thesauri et pretiosæ supellectilis, structuræ atq; picturæ.
quæ hic simul in theatro conquiri consuluntur, ut eorum
frequenti inspectione tractationeq; singularis aliqua
rerum cognitio et prudentia admiranda,
citó, facilé ac tutó comparari
possit. autore Samuele à
QVICCHEBERG BELGA.

MONACHII
Ex Officina Adami Berg typographi.
Anno M.D.LXV.
Cum gratia et priuilegio Cæsareo.

Samuel Quiccheberg's treatise on the ideal museum

According to Quiccheberg, the first collections were already described in the Old Testament (such as King Hezekiah's treasure-house and armory, Book 2, Kings 20, 12–21). He considered Solomon's temple and even the Garden of Eden to be prototypical universal collections, thereby establishing the scriptural precedents that made it desirable and proper for true Christians to also build their own natural history collections. Later writers (such as Neikelio, 1727), observed further that in building a universal collection, the curious and studious man, living in the macrocosm ordered by God, becomes similarly the creator of a parallel, ordered microcosm.

Thus developed the European notion of the universal collection, its goal being to encompass, through its examples, the full scope of the natural world, the three "Kingdoms of Nature" (plants, animals, and minerals). Such a collection was referred to as a "cabinet of curiosities" or a "closet of rarities" (in England), a *theatrum naturae* or *studiolo* (in Italy), and a *Wunderkammer* or *Naturalienkabinet* (in Germany). These latter were in the etymological tradition of the German *Kunstkammer* of "chamber of arts," and the *Schatzkammer* ("treasure chamber"), but collecting paintings, jewelry, and rare art objects remained more or less within the financial resources of only the very wealthy.

Cabinets of curiosities also often contained what were called *artificialia*, consisting mostly of ethnological artifacts, scientific instruments, and so on. Interest was engaged not only among scholars but generally among the middle class as well; even if not always prepared to study with scholarly zeal, the public was fascinated with natural history and eager to be awed by physical objects of nature.

The great appeal of the *Wunderkammer* was that it could be built on virtually any budget, containing specimens of princely extravagance or things exclusively field collected and traded. Anyone with an interest could participate, and cabinets could be made to emphasize whatever particular specialty local abundance or the collector's fancy might favor. (The situation remains the same today.)

By the end of the 16th century, the majority of mineral collectors were no longer mineral specialists but natural history generalists. Many owned minerals merely as a footnote to their more lurid and entertaining plant and animal curiosities, monstrosities, and freaks of nature. Often these collections were designed chiefly to thrill and delight casual visitors from the higher levels of society, as a means of elevating the owner to loftier social circles than he could otherwise achieve.

Thus the divisions were already drawn between the three principle types of collectors: (1) the scholars, scientists, and serious amateurs, (2) the wealthy, aristocratic collectors, and (3) the nonscientific (one might even say antiscientific) accumulators whose goals were entirely social. This is not to say that all scholars were immune to the temptation to lard their collections with grotesque rarities for social purposes. But the principle emphasis was usually clear in each case.

Following is a closer look at the activities of two of the earliest 16th-century aristocratic mineral collectors, Archduke Ferdinand of Tyrol and Emperor Rudolf II of Prague.

FERDINAND OF TYROL *(1529–1595)*

Ferdinand II was born the son of Holy Roman Emperor Ferdinand I (1503–1564) and Anne of Hungary (1503–1547). He was given a princely education by Jesuit tutors and, in 1547, at the age of 18, was installed by his father as governor of Bohemia.

Ferdinand I abdicated in 1564, passing the Imperial title to his eldest son, Maximillian II (1527–1576), Archduke of Austria and King of Hungary and Bohemia. At Ferdinand's death in 1564 his younger son, Ferdinand II, was appointed governor of Tyrol and the adjacent western territories. Spared the pressures of directing an empire, Archduke Ferdinand II immediately set about the task of rebuilding and refurbishing the family *Hochschloss*, a medieval castle dating back to the 11th century, situated in the Tyrolean mountains at Ambras.

By 1566 the initial work on the palace had been completed; in 1571 construction work began on a large banquet hall; and in 1573 the Archduke initiated

Archduke Ferdinand of Tyrol

a new building program in the Lower Castle or *Unterschloss*, a series of four large, interconnected buildings. These spacious facilities were being remodeled to house Ferdinand's collections, principally his extensive arms and armor collection, which was to have three of the four rooms. The fourth room, however, was set aside for his *Kunstkammer*.

A visitor to the archduke's museum began with the first hall of the armory (devoted to equipment for tournaments) and eventually proceeded through to the library and *Antiquarium* in an adjacent building called the *Kornschütt*. The *Kunstkammer* occupied a wing connecting the *Kornschütt* with the third hall of the armory. Before entering the *Kunstkammer*, the visitor passed through the *Türkenkammerl* or Turkish chamber, containing ethnographic objects, especially trophies from the Turkish wars.

The items in the *Kunstkammer* itself were arranged in 18 cabinets positioned back-to-back in a row down the center of the room. Facing them were two additional cabinets called *Zwerchkästen*. Wall spaces between the windows facing each side were covered by paintings; and in the center of the room were several chests containing portraits, graphic works, and items currently under study. Items in the cabinets were organized according to composition, objects made of the same material being grouped together. Thus the first cabinet contained gold objects, the second items of silver, and so on, according to Pliny's approach in stressing the connection between a work of art and the materials it is constructed from. This was also the plan followed by the Medici in their own museum or *studiolo* in Florence.

As an artistic complement, each cabinet was painted a different color so as to harmonize nicely with its contents; blue, for example, was used for the gold cabinet and green for the cabinet containing the silver items. And lighting was excellent, with a wall of windows facing each row of cabinets.

Although the *Kunstkammer* contained many objects of art, it was also well stocked with *naturalia*: minerals, plants, animals, bones, coral, sponges, reptiles, fish, and fossil dinosaur bones (thought to be the bones of "giants"), all mixed in with the other categories of man-made objects.

The archduke was not of a scholarly bent, however, and there is no evidence that he ever attempted a scholarly assessment of his specimens, nor did he keep catalogs or lists of contents. To his credit, however, he paid no attention to purported mystical properties either. He simply pursued the universal interests of a princely collector.

Although he had an appreciation for specimens in their raw form, Ferdinand was also attracted to their potential for being worked into art objects, especially those, like coral, which could retain some of their original form. He had artisans assemble miniature scenes (called "handstones") incorporating natural crystals, wire silver, and tiny figures on silver pedestals. In addition he collected *artefacta* (man-made objects), *scientifica* (scientific instruments such as astrolabes, watches, and compasses, as well as automatons), *mirabilia* (such as the largest known bowl of wood, playing cards for giants and dwarfs, illustrations showing people with gross deformities, and so on), *exotica* (ethnographic items such as pre-Columbian feather works), and one of the largest numismatic collections of his time. Some of these many items were inherited by Ferdinand from his father and brother. But many more were of his own judicious choosing.

Ferdinand had begun his collecting in Bohemia, and had also obtained specimens from the Tyrolean silver mines at Schwaz. He acquired large numbers of Milanese quartz crystals, gold specimens, over 1,200 specimens of different silver ores and minerals from various localities including South America, splendid stephanite and argentite specimens from St. Joachimsthal, cassiterite from Schlaggenwald, azurite and malachite from Schwaz, quartz from the Alps, and emeralds from "Peru" (now Colombia).

The collection stayed at Schloss Ambras for many decades following Ferdinand's death in 1595, but Emperor Rudolph II in Prague had promptly taken away many specimens, and other specimens were ultimately removed to Vienna and incorporated into the Imperial cabinet there during the following century. A large part of the mineral collection is believed to have been transferred with other *naturalia* to the University at Innsbruck in 1817.

EMPEROR RUDOLF II *(1552–1612)*

Rudolf was born in Vienna, the son of Holy Roman Emperor Maximillian II and Maria, daughter of Emperor Charles V. At the age of 11 he was sent to Spain for his education, and took easily to the scholarly environment. He showed a high level of intelligence, although he lacked the tolerant spirit of his father.

By the time he returned to Vienna he had become a cultured young man. He was interested in astronomy and astrology, later extending his patronage to support the astronomers Tycho Brahe and Johannes Kepler. Rudolf was also very much interested in minerals, chemistry, and alchemy, as well as art, and was something of a scholar himself.

Emperor Rudolph II

In 1572, at the age of 20, he was crowned King of Hungary, and three years later King of Bohemia and Germany, becoming emperor on his father's death in 1576. He was not a personable ruler, however, and became increasingly subject to attacks of depression and eccentricity bordering on insanity. He was haughty, suspicious, and reclusive, and he refused to marry.

Scholarship and culture were his refuge from the troublesome affairs of state, and he "became the greatest collector of his age, his agents ransacking Europe to fill his museums with rare works of art." He collected excellent paintings, sculpture, objects of Western craftsmanship, a large number of everyday objects from the Far East and Middle East as well as the Americas, and *naturalia* including hummingbirds, crabs, lobsters, shells, a mammoth tusk, and fossils. He pursued minerals and gemstones with great enthusiasm, building an enormous collection.

Rudolf settled in Prague in the 1580s, at Hradschin Castle. Work began in 1587 on a summer palace, which served as a temporary home for his growing collections. In 1597 the collections were being housed in three rooms of the southern wing, but they were moved into new rooms in 1605 or 1606 when construction was finally completed. The *naturalia* were housed in the *Kunstkammer*, consisting of three vaulted chambers in front, each about 5.5 meters wide and 3 meters high and all together about 60 meters long, connected to the principal chamber, 33 meters long. The *Kunstkammer* contained 37 cabinets. "Large natural gemstones" (uncut crystals of gem minerals) were kept in several chests.

Rudolf's *Kunstkammer* was not a mere cabinet of curiosities but rather a *systematic* collection of objects and specimens from the three Kingdoms of Nature and of the works of man, based on an encyclopedic principle. It was not designed and housed to be seen by a wider public, but only for those few initiates whom Rudolf admitted to his inner circle. It also represented an important source of inspiration and raw materials for court artists.

Rudolf employed his court physician, Anselmus Boetius de Boodt (ca. 1550–1632), to curate his specimens. DeBoodt had studied medicine at Heidelberg and obtained his medical degree in Padua. From 1583 he lived in Bohemia as physician to the burgrave of Prague, and was a close friend and collecting partner of the Imperial physician Thadeus Hayek, a well-known naturalist. DeBoodt was a polyglot, an artist, and an enthusiastic mineral collector. He made many mineralogical field trips into Germany, Silesia, and Bohemia collecting specimens. With Rudolf's vast collection at his disposal, he went on to write one of the most influential mineralogical works of the 17th century, *Gemmarum et Lapidum Historia* (1609).

Because of his increasing psychological instability, Rudolf's relatives had him declared incapable of ruling in 1606, and placed his younger brother Matthias more or less in charge. Various difficult intrigues and political unrest followed; in 1612 Rudolf died, to be officially succeeded by Matthias as emperor.

Rudolf's mineral collection was not appreciated by his successors. It was eventually placed into storage, and in 1782 was packed in barrels and sold off to private parties by Joseph II.

The 17TH CENTURY

During the greater part of the 17th century, the trend in collecting continued to depart (one might say regress) from the specialized mineralogical collections pioneered by the founding fathers of mineral collecting such as Giorgius Agricola, Johannes Kentmann, and Michaeli Mercati. Nearly all collections were of general natural history, including plants and animals, and many also included artworks and other *artificialia*. Furthermore, a degree of intellectual vitality was lacking in many of them.

With regard to the Habsburg collections, Distelberger (1985) writes:

> *So completely do the great collectors Archduke Ferdinand II (1529–1595) and Emperor Rudolf II (1552–1612) overshadow the succeeding generation [of Habsburg collectors], that the seventeenth century came to be seen mainly as a time of consolidation rather than of acquisition, and it is certainly true that the Thirty Years War no longer permitted so intensive a dedication [to collecting] as that which Rudolf II had shown.*

The number of minerals in the *Kunstkammer* of the Saxon electors continued to grow, and the collections of the Italian princes were maintained and enlarged. But the Habsburg collections were in decline, and important princely collections were to change in their basic concept and orientation. Francesco de Medici, for example, reorganized the private family *studiolo*, opening the Uffizi Gallery to the public in a politically oriented attempt to glorify and legitimize his dynasty.

Works of art gradually became more separated from *naturalia* of scientific value. Among the private collections of Italian physicians and naturalists there was a distinct movement toward purely natural history collections, but these collectors failed to gain the widespread fame that the museums of Ulisse Aldrovandi and Francesco Calzolari had enjoyed a century earlier. Olmi (1985) attributes this drift to the greatly increased availability of mineral, plant, and animal specimens from distant lands. This abundance deprived many collections of their novelty and exclusivity, making them less interesting to the public. Many non-scholarly collectors, especially in Italy, whose goals had been primarily social, abandoned the pursuit of exotic *naturalia* and returned to art collecting as the best means of enhancing their social standing.

Encyclopedic collections combining all types of natural and artificial specimens thus declined somewhat. However, some excellent collections in northern Italy, for example, continued to be all-inclusive, notably those of Manfredo Settala, Lodovico Moscardo, and Ferdinando Cospi. But unlike their scholarly predecessors, these collectors lacked college education and did not focus on scientific research. Even the scientific instruments in their collections were seen not as investigative tools but rather as precious, aesthetic relics. Their natural history specimens were "organized in such a way as to exclude systematically all normality"; they were focused on the bizarre curiosities most likely to provoke astonishment and wonder.

As the 17th century progressed, a growing number of collectors and curious-minded scholars traveled about Europe on a regular round of visits to collections. They were aided by various printed itineraries and guidebooks. Although collections of the most broad and general scope remained common, many others were more specialized, several hundred of which were devoted solely to natural history. Some even maintained regular hours and charged a small admission fee. The exchange and purchase of specimens continued at a brisk rate, some collectors being more generous and benevolent to their colleagues, and others being more mercenary. One prominent traveler and writer, Zacharias Conrad von Uffenbach, commented as follows on the British collector James Petiver (1663–1718) (whose minerals and other curiosities were purchased later by Sir Hans Sloane): "He offers all foreigners who come to him a sample of his collection; but he takes care to ask a vast sum for it, so I declined with thanks."

In England, natural history cabinets containing minerals developed on a smaller scale than in Italy. These collections were closely modeled on their counterparts in Italy and Germany, but did not begin appearing until the 17th century. The collections of naturalists such as the physician William Harvey (1578–1657) and John Tradescant (1587–1638) contained minerals, and the encyclopedic London museum of William Courten (1642–1702) contained over 10,000 cataloged mineral specimens.

Sir Hans Sloane (1660–1753) surely ranks as England's greatest collector of natural history; he acquired the Courten collection among many others, and eventually donated all of his specimens (along with a scientific library of 50,000 volumes) to the nation in order to found the British Museum.

The first mineral collector of record in what was to become the United States was John Winthrop the Younger (1606–1676), son of the first governor of the Massachusetts Bay Colony (John Winthrop, 1588–1649), and himself a colonial governor. He had been born in England and educated at Trinity College, Dublin, presumably acquiring there some interest in natural history. He possessed a good library and was considered to be the "leading scientific mind in the colonies." As leader of a group that settled Ipswich, Massachusetts, in 1633, he was later appointed governor of a vast plantation in Connecticut and then of the entire colony (1657, 1659–1676). Winthrop owned mining properties in Massachusetts and Connecticut; and in 1647, in Saugus, Massachusetts, he began the first commercial iron mining and smelting operation in the colonies. He appears to have been an active prospector, having made the first discovery of gold in New England (Great Hill at Cobalt, south of Portland, Connecticut). His mineral collection of New England specimens did not survive, but his catalog did; it was published, in part, in the *Journal Book of the Royal Society* by his grandson, also named John Winthrop, in 1734. The article was entitled "Selections from an ancient catalog of objects of natural history formed in New England more than 100 years ago."

By the end of the 17th century, encyclopedic collections were beginning to come under serious criticism. Count Luigi Ferdinand Marsili (1658–1730), an Italian naturalist with a more specialized collecting approach, severely condemns the mixing of *naturalia* and *artificialia*; such mixed collections, he asserts, serve "more to delight young men and provoke the admiration of women and the ignorant" than to serve any legitimate scholarly purpose (Spallanzani, 1987). To Marsili, even the most common specimens of minerals, plants, and animals had a necessary place and purpose in a truly scientific collection. This concept, shared by John Woodward, was a revolutionary idea in collecting: to acquire the mundane and superficially uninteresting along with the spectacular for the sake of systematic, scientific completeness. Such collections, particularly those specializing in minerals, were to become ever more common in the following century as the scientific revolution took hold.

The 18TH CENTURY

In the opening years of the 18th century two important reference works for collectors appeared that more or less summarized for the layman the collecting scene at that time. These books, by Michael Valentini (in 1704–1714) and Caspar Neickelio (in 1727), list collections in various countries, discuss their contents, and review collecting philosophy and recommendations.

MUSEOGRAPHIA
Oder
Anleitung
Zum rechten Begriff und nützlicher Anlegung
der
MVSEORVM,
Oder
Raritäten-Kammern,
Darinnen gehandelt wird
I. Von denen Museis, Schatz- Kunst- und Raritäten-Kammern insgemein, welche heutiges Tages grösten theils annoch in vielen Europæischen Orten gefunden werden.
II. Dem nachmals ein Anhang beygefüget ist, von vielen, welche vor Alters in der Welt berühmt gewesen.
III. Im dritten Theile wird von Bibliothequen insgemein, als einem zu einem vollständigen und wohl eingerichteten Museo unentbehrlichen Wercke gehandelt.
IV. Der vierte und letzte Theil aber ist eine Anmerckung oder unvorgreiffliches Bedencken von Raritäten-Kammern oder Museis insgemein.
In beliebter Kürtze zusammen getragen und curiösen Gemüthern dargestellet
von
C. F. NEICKELIO.
Auf Verlangen
mit einigen Zusätzen und dreyfachem Anhang vermehret
von
D. Johann Kanold,
der Kayserl. Leopold. Carol. Reichs-Academiæ Natur. Curios. Mit-Glied.
Nebst einem Register.
Leipzig und Breßlau, bey Michael Hubert, 1727.

Michael Valentini's book of museums

Valentini's book, *Museum Museorum*, reprints in its entirety the text of an earlier museological tract of 1674 by Johann Daniel Major (1636–1693). Major discusses the ways in which general collections (*naturalia* and *artificialia*) can be arranged and addresses the basic question of why people collect. He lists the important collections known to him and gives practical advice on specimen conservation and on the study of museum materials. He recommends in particular that a collector prepare a catalog or "rarities book" to accompany his collection, describing not only his own specimens but also all other things that would be necessary to fill it out to optimal completeness. Major's work was the first widely circulated essay defining the various types of collections, and by 1700, preparing their own "rarities book" had become customary among collectors.

In the remainder of his large two-volume tract, Valentini concentrates primarily on addressing collections of natural history. He lists many collections in different cities, often giving detailed lists and descriptions of their contents. His own collection is among those featured. He also provides a general text on mineralogy, botany, and zoology for collectors to refer to, illustrated by engravings.

Caspar Neickelio, in his *Museographia* (1727), presents a geographically arranged compendium of all the known collections, giving little or no detail on contents in most cases (with a few notable exceptions). He also lists all of the important libraries of the time, and discusses numerous practical and philosophical questions of interest to the collector. Neickelio was a strong believer in specialization, or at least in compartmentalization, and he describes numerous fields on which a collector might focus. In the preparation of a mineral collection, for example, he recommends that it include the following:

> *All kind of Minerals (the Mineral Kingdom), for example all kinds of stones;* gems *such as diamond, ruby, sapphire, emerald, etc.;* rarities *such as marble, bloodstone, osteocolla, etc.;* common *things such as sandstone and conglomerate, etc. One also has the stones found in some animals such as crabs, carp,*

> *toads, old roosters, and indeed even in some men (which according to the best authorities belong to the class* animalia*). To the Mineral Kingdom belong all sorts of* metals: *gold, silver, copper, tin, iron, lead, mercury or quicksilver, etc.; materials that are* semi-metalla *or* mineralia *such as cinnabar, antimony, etc.* Recrementa Metallica, *or unripe metals, such as cobalt, marcasite, chalcitis, iron sulfates, porri, etc.* Salia, *all kinds of salts, such as saltpeter, common or cooking salt, vitriol, alum, etc.* Sulphurea, *sulfurated minerals, arsenic, sulfur, amber, naphtha, asphaltum, etc.* Terrestria, *such as earths, clays, marga, tripoli, chalk, etc. And, in short, all kinds of minerals and fossils that are found under the earth.*

Having informed his readers of the whereabouts of hundreds of private collections and museums of all sorts which they might conceivably be able to visit, Neickelio then thoughtfully includes a list of 25 guidelines to follow for proper etiquette and maximum benefit when visiting someone's collection.

Regarding the composition and arrangement of an ideal private museum, Neickelio adheres to the old idea that it should be all-inclusive, a universe in miniature, but he also maintains that the different categories of specimens should be well separated and segregated from each other for conceptual purposes. Ideally, in addition to the three basic natural history sub-collections (animals, plants, minerals), a comprehensive selection of artworks should also be included ... man's creations to be exhibited beside those of God. A study of Vitruvius's classical treatise *De architectura* led Neickelio to further recommend the specific location of the collection rooms within a building, that the rooms be well ventilated, that a museum room be roughly twice as long as it is broad, that it be oriented in a southeasterly direction, the walls painted in a light color, with a central entrance and a large work table. Bookshelves should be installed on one side of the room (a collector's library being very nearly as important as the specimens themselves), with sets of shelves for specimens on the facing wall, separated by windows. He specifies that the arrangement of the specimens should be orderly, not only grouping all the minerals together, for example, but also with the larger objects on the lower shelves and the smaller specimens higher up. To remind visitors of the silence necessary for study, a fearsome pair of stuffed or sculpted lions, bears, or tigers should flank the entrance. A register of new acquisitions, a complete written inventory, and a method for conveniently accessing items are also required.

Neickelio's Ideal Museum

With such practical, helpful reference works as guides, scientific and aesthetic mineral collecting as a common specialization eventually burst into full bloom in the latter half of the 18th century. The excellent beginning established for specialized mineral collecting in the 1500s was followed in the next century by a conceptual decline and dilution, as the emphasis shifted almost exclusively to broad-based, all-inclusive natural history and ethnographic collections. But this trend was reversed during the course of the 18th century, as scientific understanding of minerals developed more fully. General, mixed collections of *naturalia* and *artificialia* became more sorted and compartmentalized according to the recommendations of Valentini and Neickelio. Specialists again emerged, in the mold of Agricola, Kentmann, Mercati, and Palissy, and they leaned heavily for guidance on the writings of the earliest collectors. At the same time, collectors with a new aesthetic sensibility toward minerals began to develop a kind of sophistication unknown in the 16th century, resulting in the formation of some extraordinarily beautiful assemblages of minerals. Thus the modern concept of connoisseurship encompassing both beauty and science had at last begun to evolve.

Mineral collections of the 18th century were able to draw upon a much expanded system of sources and an active specimen market fed by many professional mineral dealers. Mineral enthusiasts with sufficient funds could now build enormous and expensive collections without having to field-collect so much of their material, a situation that served to promote the concept of connoisseurship, as well as to establish significant monetary values for the best specimens and the finest collections.

John Stuart, the Earl of Bute

Although the majority of 18th-century mineral collections were probably relatively modest in size, containing perhaps 200 to 1,000 specimens, others were very large indeed. The largest private mineral collection was reputed to be that of John Stuart (1713–1792), the Earl of Bute, who is said to have had as many as 100,000 specimens. No catalog of the Bute specimens giving a count survives, and the figure is probably inflated. But a significant number of other collections containing 10,000 to 20,000 specimens is known to have existed, and several were attested to be in the 30,000 to 40,000 range. Of course, a large count is no guarantee of specimen quality, and some large collections (e.g., those of Franz von Beroldingen and Hans Sloane) are considered to have been of mediocre quality. But others, such as those of George Gibbs, Charles Greville, and Alexander Crichton, were assembled according to the highest standards. Unfortunately, the sizes of the great majority of mineral collections of the time are unknown; but a review of the largest collections of established size demonstrates how substantial many of these collections became.

In most cases (unfortunately including many of the most important transactions) the selling prices of collections made or begun in the 18th century also remain unknown. However, some notable exceptions exist that serve to give some idea of the value that had come to be accepted for large, excellent collections of minerals. Alexander Crichton's collection of around 4,000 superb specimens sold for the equivalent of about $2.3 million in 1992 dollars. Pedro Davilla's collection sold for $3 million; the collection of the pioneering crystallographer René Haüy sold for $3.2 million; and the collection of the Princes Lobkowitz sold for $6.7 million (all 1992 dollars).

Selling prices can be informative in a broad sense but must be viewed with some caution, because significant underpayment and overpayment were common. Furthermore, it can be difficult to establish precise equivalents of

18th-century currencies with modern money. However, because all European currencies were at that time based on gold, they can at least all be reduced to equivalent gold weights. And an approximate modern equivalent value of gold per ounce can be cautiously postulated on the basis of gross economic comparisons. Even given the uncertainties in the calculation, there is no doubt that collections that sold for upward of half a million 1992 dollars must have been extraordinary, and were also surprisingly numerous, at least seven having changed hands for the equivalent of over a million dollars. This is a phenomenon that has probably happened in America only twice in the 20th century (the 1984 sale of the Perkins Sams collection to the Houston Museum of Natural Science and the recent sale of the William Pinch collection to the Canadian Museum of Science).

Many interesting examples of aristocratic mineral collection in the 18th century could be cited. The following discussion is arranged by country.

FRANCE

Royal enthusiasm for mineralogy in France was an inspiration to collectors and mineralogists throughout the 18th century. It began with Louis XIV (1638–1715), who in 1671 issued an edict regulating the *Jardine du Roi* ("King's Garden") and the royal pharmacy where medicinal plant and mineral specimens were kept. These officially became the *Cabinet du Roi* ("King's Cabinet") upon his death in 1715; under the regency of Philip of Orleans (1674–1723) it was organized by Bernard Jussieu (1699–1777) along the lines of the great 17th-century Italian collections.

Louis XV (1710–1774), who was only five years old when his father died, took an interest in the collection. He appointed the famous naturalist Georges Buffon (1707–1778) curator in 1739, and permitted Buffon to open the cabinet to the public in 1745. It thus became the first publicly accessible museum in Paris (after that of Bernard Palissy many years before), opened five years before the Louvre. Louis Daubenton, Buffon's deputy, took over primary responsibility for the mineralogical portion of the cabinet.

Louis XVI (1754–1793) retained Buffon and Daubenton after his accession to the throne in 1774, and continued to assist in building the collection. Having married the daughter of Austrian Emperor Francis I and Empress Maria Theresa, enthusiastic mineral collectors themselves, he received many mineral specimens from the Imperial Cabinet in Vienna. Important donations were also received from Catherine the Great of Russia (1729–1796), King Christian VII of Denmark (1749–1808), and various traveling naturalists. The private collections of Réné Réaumur (1683–1757) and Joseph Bonnier de la Mosson (1702–1744) were acquired in 1776–1779. Joseph Dombey (1742–1793) sent many specimens collected during his travels in Mexico and South America in 1779; many of these, including atacamite and euclase, became important research specimens. Déodat

Dolomieu (1750–1801) sent zeolites from the Cyclopean Islands, and sulfur with celestine from Sicily. By the time of the French Revolution, the king's cabinet had grown to include about 2,600 mineral specimens.

The king's cabinet was taken over by the new government of the revolution, and was re-established in 1793 as the Museum of Natural History. Daubenton became the first professor of mineralogy at the new institution, to be succeeded in due course by Dolomieu and Haüy.

Many of the most important private mineral collections of 18th-century France were, in time, acquired by the Natural History Museum. Balthazar Sage's collection was divided in 1825 between the museum and the Paris School of Mines, which he had founded. Romé de l'Isle's important collection had been acquired, following his death in 1790, by Gillet de Laumont, whose 1,600 specimens (including de l'Isle's) were purchased by the museum in 1835. Haüy's mineral collection, sold by his niece to the Duke of Buckingham in 1823, was purchased back from the duke's heirs in 1848. At the same time, many of Europe's most prominent mineralogists and chemists, who had also begun their collections in the late 18th century, made frequent donations of mineral specimens. Minerals were received from naturalists such as Karl Ludwig Giesecke (1761–1833), Louis Leschenault de la Tour (1773–1826), Pierre Marcel de Serres (1780–1862), Louis Vauquelin, and many others.

Following the restoration of the Bourbons in 1814, Louis XVIII (1755–1824) welcomed his friend, the loyal expatriate Count de Bournon, back from voluntary exile in England. Louis had no doubt enjoyed the mineral cabinet of his brother, Louis XVI, prior to the revolution. He had spent time with Bournon in England during the Napoleonic wars, and had studied mineralogy under Bournon's guidance. Back in Paris, Louis purchased Bournon's collection of nearly 23,000 specimens and installed Bournon as its new curator. Thus, for a few years, the King of France once more had a private mineral collection.

Bournon and Louis XVIII died in the same year, and the Natural History Museum was given most of Louis's mineral collection.

AUSTRIA

In large part inspired by the enthusiasm of the Royal Imperial family (Emperor Francis I [Franz] and his wife Maria Theresa), the city of Vienna became one of the great hotbeds of European mineral collecting in the 18th century. Aristocrats, scholars, the clergy, military men, businessmen, and medical men – all catered to by a remarkable number of full-time mineral dealers – comprised a very active, enthusiastic, and competitive mineralogical community. Scholars from all over the empire were called to Vienna and put to work tending the Royal Imperial collections, teaching in the Imperial academies, and serving in a plethora of official posts. Many of these scholars were, or became, mineral collectors in their own right.

Over the course of the 18th century, the love of mineral collecting spread downward through the Imperial family. Archduchess Maria Anna (1738–1789), daughter of Franz and Maria Theresa, accumulated nearly 10,000 mineral specimens in her personal collection. Their sons Joseph (1741–1790) and Leopold (1747–1792), who each in turn inherited the Imperial title along with the Medicis's collection of natural history, were less interested, but Leopold did found the Imperial Museum in Florence where he put the Medici minerals on public display. Archdukes Johann (1782–1859) and Rainer (1783–1839), sons of Leopold, each built significant mineral collections of their own while still young, and purchased other collections as well. Johann's extensive collection eventually helped to found the Steiermark Provincial Museum, known affectionately in his honor as the *Joanneum*.

Other Austrian aristocrats known to have collected minerals include Princes Johann (1760–1836) and Louis (1780–1833) von Liechtenstein, the latter of whom employed the famous mineralogist Friedrich Mohs as his curator and purchased other collections (including that of Count Leopold von Kolowrath-Krakowski). Prince Ferdinand (1763–1834), Duke of Württemburg and Commandant of the city of Vienna, traveled on collecting trips to Russia where he acquired many beautiful Russian and Siberian mineral specimens; he later donated his collection to the Imperial Cabinet. Among the most important Austrian mineral collectors was Prince Prosper von Sinzendorf (1773–1822); Imperial Curator Andreas Stütz ranked Sinzendorf's collection as equaling or surpassing most of the other collections in Vienna, and lamented the prince's removal of it to his castle in Ernstbrunn.

Count Leopold Kolowrath-Krakowski

Count Rudolph Wrbna (1761–1823), President of the Royal Imperial *Bergkammer*, built a costly mineral collection of his own, which was rich in Hungarian specimens. Count Ferdinand von Palfy (1774–1840) built another collection at great cost, traveling personally to buy mineral specimens in Hungary and other countries. Count Dietrichstein based his collection in Vienna on the minerals acquired from Abbot Anton Estner (1730–1801), curator of Count Wrbna's collection. Baron Thaddaeus von Lichtenfels (1764–1829), Baron Michael von Kienmayer (1755–1828), and the Marquis von Malaspina are also known to have built mineral collections.

In fact, a large percentage of the Austrian nobility probably at least dabbled in mineral collecting at one time or another. Baron Ignaz von Born (1777) writes approvingly of the inspiration young aristocrats were given in school:

> *The [mineral] collections of the [Viennese] academies for noblemen have afforded me great delight, not indeed for their rarities, but by their inspiring young noblemen with some relish for this science. Every nobleman in the Theresian college [founded by Maria Theresa] has a small cabinet of minerals, shells and insects in his apartment.*

Emperor Franz Stefan von Lothringen

Among the most important men in the history of Austrian mineral collecting was Franz Stefan von Lothringen (1708–1765). The head of the Royal Family in Vienna, he was the son of Leopold, Duke of Lorraine (Lothringen). Franz succeeded to the duchy in 1729, but ceded it to the king of Poland in 1737. The year before he had wed Maria Theresa of Austria, and he ruled Austria with her as co-regent from 1740 to 1745, when he was chosen as Holy Roman Emperor. He did not concern himself much with the wars that were going on, but preferred to indulge in his favorite pastime: collecting minerals and other *naturalia*.

In 1748 Franz (now Francis I) purchased the enormous natural history collection of a Florentine scientist, Ritter (Sir) Johann von Baillou (1679–1758). This collection, which contained over 30,000 minerals, rocks, and shells, was at that time one of the world's largest private collections of such objects. Franz had it brought to Vienna and installed it in the Hofburg palace as the Imperial private collection; von Baillou was retained as curator.

Franz was an amateur scientist at heart, with a partiality for medicine and chemistry and a fondness for physical and mechanical scientific instruments and models. He even pioneered in the manufacture of some of his own instruments, including a large concave mirror with which he and the Jesuit Father Joseph Franz succeeded in demonstrating the combustibility of diamond in 1751.

In 1755 Franz commissioned the botanist Baron von Jacquin to undertake a collecting expedition to the Antilles, Venezuela, and Colombia. The trip, which lasted four years, was the first officially sponsored overseas expedition on behalf of the Imperial Cabinet. Fifty large crates were shipped back to Vienna; included were zoological specimens, plant specimens, ethnographic items, and many valuable minerals.

To Franz, however, all of his collecting was for the purpose of his personal enjoyment only, an attitude not shared by his wife. It was not until Franz's death in 1765 that she was able to share his magnificent collection with a wider audience.

Maria Theresa, Archduchess of Austria and Queen of Hungary and Bohemia, was the daughter of Charles VI (1711–1740), Holy Roman Emperor, King of Hungary, and last of the male line of the Habsburgs. Maria Theresa apparently took some interest in Franz's collecting activities. Unlike her late husband, however, she did not regard the collections as personal playthings but as national treasures of some potential use in educating and elevating the public. Therefore she formally presented his collections to the Austrian state and established for them a public museum (among the first in history) known as the Imperial *Naturalienkammer*.

It was also her motive that, in addition to providing pleasure to the public, the exhibits should spread the knowledge of the appearance of economically valuable minerals more widely, facilitating new discoveries. She later called the famous mineralogist Ignaz von Born to Vienna to curate the natural history collection, and charged him with organizing it according to the latest scientific knowledge, and seeing that all Imperial mining properties contributed representative specimens.

The Imperial collection grew rapidly through donations and purchases from private collectors, including Bergrat Leithner in Idria; professor Anton Ruprecht (1748–1814) at the Schemnitz Mining Academy; Baron Franz Müller von Reichenstein (1740–1825, discoverer of tellurium), who was the director of all mines in Transylvania; Bergrat Karl von Ployer (1739–1812) in Carinthia; and the famous mineral collector Johann Fichtel (1732–1795); among many others.

Today Franz Stefan's minerals and all that was later added to his collection are still preserved in the Vienna Museum of Natural History.

HUNGARY

Georgius Agricola had written much about the mines of Upper and Lower Hungary in his *De Re Metallica* (1556), tantalizing later mineralogists to venture into those lesser known parts of Europe. Like the Saxon Erzgebirge, the metalliferous mines of Hungary yielded large quantities of silver, and also gold, copper, and iron. These mines, because of their richness and longevity, became a second major focal point of mining culture in Europe, spawning, as the Saxon mines had done, a thriving local enthusiasm for mineral collecting.

There was evidence for this enthusiasm even in the 17th century. Emperor Rudolf II (1552–1612), for example, ordered the magistrates of Schemnitz and Kremnitz and the Bishop of Eger to send him fine mineral specimens "from their collections," attesting to the accumulation of private Hungarian collections even at that early time. When Edward Browne (1644–1708), personal physician to the English King Charles II, visited Hungary in 1669–1670 he found many people collecting minerals there. (Perhaps the most notable of these was M. Dollinger, the director of mines in Kremnitz, but there were many others.) Hungarian mineral wealth became ever more famous throughout Europe.

In the 18th century, the German physician E. F. Brückmann traveled through upper Hungary in 1724, and later wrote of having seen many fine mineral collections. The French Jesuit F. X. Feller wrote of his visit to Hungary in the late 1760s that "the Hungarians have a special liking for natural history. Collections of minerals and other rare and valuable objects can be seen at many places" (Papp *et al.*, 1991). Ignaz von Born (1742–1791), who was born in Transylvania but spent much of his life in Prague and Vienna, traveled there as well in 1770. And the British scholar Daniel Clark followed his trail in 1802, both of them witness to active mineral collecting and dealing by commoners and aristocrats alike.

Unfortunately, many Hungarian collections were confiscated by the Habsburg rulers and removed to the capitals of Vienna or Prague. Such was the fate of Count Ferenc Nádasdy (1625–1671) and his extensive treasury, including minerals; he was caught in a plot to overthrow the Habsburgs and was executed in Vienna.

But during the 18th century other important collections were built that fared much better. Count Ignac Batthyány (1741–1798), a Transylvanian bishop, built an enormous collection of minerals, fossils, shells, antiquities, and books, which he ultimately established as a museum called the "Batthyáneum." Along with the collections of the Barons Bruckenthal (Samuel, 1721–1803; Michael, 1746–1813) in Transylvania, Prince Josef Lobkowitz (1772–1816), the Counts Andrássy, and the counts, dukes, and princes of the Esterhazy family, it formed the basis for some of the most important public collections in later years. Other aristocrats such as Countess Julianna Széchény (1753–1824) and the Counts Teleki (Samuel, 1739–1822; Domokos, 1773–1798) also helped to stock public collections through the donation or sale of their collections.

Prince Lobkowitz's collection (expanded significantly by his son, Prince Ferdinand Lobkowitz, 1797–1868) was "probably the greatest and most impressive private geological collection" in Austria-Hungary. It included a large systematic mineral collection (11,675 specimens), a smaller traveling collection (2,962 specimens), a collection of minerals from his own estates (2,492 specimens), and other rocks, fossils, and ornamental stones, bringing the total to over 40,000 pieces. It was inherited by Ferdinand's daughter, Countess Bossi-Fedrigotti, who put it up for sale. Although the Imperial Mineral Cabinet in Vienna and the Serbian government were interested in acquiring the collection, it was ultimately purchased by the Hungarian National Museum in 1870 for 30,000 florins. The collection had cost the prince more than 100,000 florins to build, ranking it as the most expensive mineral collection of the time. Much of it survives today, despite partial destruction by Soviet shelling in 1956.

BOHEMIA

The rich history of Bohemian mineral collecting goes back to Agricola himself, who served as physician to the famous mining town of St. Joachimsthal (modern-day Jachymov) from 1527 to 1533. These are the very years in which he was beginning to build the first known mineral collection; and the St. Joachimsthal mines are the setting for his *Bermannus* (1530), in which the process of mineral collecting was first discussed. Another of the very first collectors, Johannes Mathesius (1504–1565), served as a clergyman in the same city beginning in 1542. And, among the early nobility, Emperor Rudolf II (1552–1612) in Prague assembled one of the earliest and most influential mineral collections of his time. So it can be said that Bohemia was among the first centers of mineral collecting.

In the 18th century, one of the most important mineral collectors in Bohemia was Count Joachim von Sternberg (1755–1808), whose extensive collections were later donated to found the National Museum in Prague. Count Frantisek Kinsky (1739–1805) donated his excellent mineral collection to Prague University in 1775, where it became the basis of the "Museum Naturae Prague."

This museum, which was opened to the public on Tuesdays and Thursdays, was curated by the university professors of mineralogy.

Other aristocratic Bohemian mineral collectors of the late 18th century included Count Anton Waldstein in Duchov (Dux), western Bohemia; Count Franz de Paula Hartig (1758–1797), Lord of Niemes and Wartenberg and a patron of science; Count Joseph Millessimo (died 1825) in Prague; Count Eusebius Pötting (died 1825?) in Prague; Count Leopold Kolowrat-Krakowski (1727–1809), whose mineral collection was sold to Prince von Liechtenstein; Count Franz Anton von Kolowrat-Krakowski (1778–1861), whose mineral collection was acquired by the Narodní Museum; Count Johann Chotek (died 1824), whose collection numbered 4,000 mineral specimens; Count Nostitz-Rhinecku (also acquired by the Narodní Museum); and Count Josef von Auersperg (1767–1829), who built yet another "beautiful collection of minerals" that was ultimately acquired by the Narodní Museum. The Moravians Count Johann Nepomuk von Mitrovsky (1757–1799) and Count Lichtenstein both donated their mineral collections to the Scientific Society of Bohemia, collections that later became the basis for the county museums in Opava, Olomouc, and Brno. Other collections were less fortunate, such as that of Count Frantisek Desfours-Walderode, destroyed by bombing in World War II.

Among the aristocratic politicians was Prince Klemenz Wenzel Nepomuk Lothar von Metternich-Winneburg (1775–1859). Although his official duties took him to many European capitals, he maintained a castle at Kynzvart in western Bohemia where he accumulated quite a substantial mineral collection. He absorbed the collections of many other people, including that of Karl Huss (1761–1839), official executioner for the town of Eger. Metternich's minerals were finally acquired by the Narodní Museum in 1945.

ITALY

The long tradition of collecting in Italy continued into the 18th century, among a few aristocrats and, more important, within the academic community. Cosimo de Medici (1642–1723), Grand Duke of Tuscany, had for many years promoted natural science and enlarged the family cabinet of curiosities with many "natural productions." When the last of the male Medici line died in 1737, their natural history collections (including minerals) were curated for a time by Giovanni de Baillou (1679–1758) in Florence; ultimately they were turned over to the public when the Imperial Museum of Florence was founded by Emperor Josef II (1741–1790). The minerals are preserved today in the collection of the Museum of Mineralogy of the University of Florence. Baillou's own collection, among the largest of his time (about 30,000 specimens), was purchased by Emperor Francis I in Vienna, and is now preserved in the Natural History Museum in Vienna.

Another important aristocratic collection was formed by Ferdinand I (1751–1825), Bourbon King of the Two Sicilies and King of Naples. His private mineral collection was donated to help establish the Royal Mineralogical Museum in Naples in 1801, and is kept today in the Institute of Mineralogy at the University of Naples.

Other highborn collectors included Prince Biscari of Florence (who published a catalog of his natural history collection in 1777), and the Duchess of Calabrito, Petronilla von Ligniville in Naples. She purchased the mineral collection of Josef van Dam in Vienna and donated specimens from Vesuvius and Sicily to the Imperial Cabinet in Vienna.

RUSSIA

The Russian nobility, like noblemen elsewhere in Europe, were attracted by the Baroque fashion for collection building. The earliest known Russian *Kunstkammern* belonged to Czar Ivan the Terrible (1530–1584) and later to Czar Alexis (1629–1676), father of Peter the Great, although these are not definitely known to have contained mineral specimens.

Czar Peter Alekseevich the Great

Peter the Great (1672–1725) was born to Natalia Kirillovna Naryshkina, a woman whose former guardian had raised her in an atmosphere open to the progressive cultural influences of Western Europe. As a child he became acquainted with residents of the nearby "German settlement," where foreigners were permitted, and became interested in the lives of other nations.

In 1697–1698 Peter traveled through Western Europe with his so-called "Grand Embassy" of 250 diplomats and scholars, examining the international situation and gathering knowledge of the intellectual and cultural life of the West. He was enormously stimulated by this exposure to foreign ways and as a result brought new vigor to the building of his own collection of rarities. He installed his first collection of *naturalia* in the Imperial pharmacy, with his personal physician, Aryeskin, as its curator.

By 1714 Peter's natural history collection had become substantial, and, in the enlightened interest of fostering public education, he opened it to the public. During his second trip through Germany, Austria, the Netherlands, and England in 1716–1717 he purchased entire collections to bring back to St. Petersburg. The most notable acquisitions were the enormous natural history collections of the Amsterdam apothecary Albertus Seba (1665–1736) and the Dutch anatomist Friderich Ruysch (1638–1731), both of which contained some minerals, and especially the collection of nearly 1,200 mineral specimens owned by J. Leopold Gottwald, a Danish physician in Danzig. A few years later, in 1724, Peter placed his scientific collections under the administration of the St. Petersburg Academy of Sciences. Most of Peter's specimens survive today as part of the 120,000-specimen gem and mineral collection preserved in the Fersman Mineralogical Museum, Moscow.

Empress Catherine the Great deposed her husband Peter III in 1762, with the help of her paramour, Grigori Orlov, Grigori Potemkin, and Princess

Princess Catherine Romanovna Dashkova

Catherine Dashkova. Both Catherine the Great and Catherine Dashkova were mineral collectors when not involved in palace intrigues. The former purchased a valuable mineral collection from the prominent German naturalist and traveler Peter Simon Pallas (1741–1811) and kept a cabinet of minerals at her winter palace. She appointed Catherine Dashkova Director of the St. Petersburg Academy of Science in 1782, and of the Russian Academy of Science in 1783, positions that Catherine Dashkova filled with distinction. Dashkova donated her personal mineral collection to the St. Petersburg Academy.

Catherine the Great's granddaughter Alexandra (1783–1801), who married Austrian Archduke Joseph, was an enthusiastic mineral collector as well. She had a systematic mineral collection of 514 specimens (317 of which came from Russia), representing 250 localities. This collection had been assembled and organized for her by Pavel Olorovskiy. In addition, she maintained a 255-specimen collection of mainly Uralian specimens. Following her death, these collections were given to the Royal Hungarian University in Budapest (now Eötvös Lorand University), where they have been preserved, at least in part, along with her two-volume catalog.

Another notable Russian mineral collector was Count Alexis Golovkin in Moscow. He was a friend and patron of the French mineral illustrator Swebach Desfontaines. He purchased several thousand mineral specimens from the collection of the French statesman Charles Calonne in 1801, at what was said to be a great price. Calonne was a prominent collector in Paris, and had sold a substantial mineral collection to Balthazar Sage in 1789, which became the foundation for the Paris School of Mines collection. When Count Golovkin's collection was dispersed following his death, some specimens were acquired by Alexander von Crichton.

Yet another prominent Russian collector was Count Aleksandr Stroganov (1733–1811), who owned salt mines and iron mines in the Ural Mountains. He built an extensive scientific library at Stroganov Palace, with a large adjoining room for his mineral collection, arranged in glass display cases. He possessed especially fine specimens of native gold from the Urals and an excellent Russian rubellite crystal group. It was preserved for some years following the death of the count, but today only the empty glass cases remain in the palace; the minerals are gone.

One of Russia's greatest private mineral collectors was Count Nikolai Rumiantsev (1754–1826). His private museum of books, manuscripts, coins, and ethnographic items also included nearly 13,000 mineral specimens. He bequeathed his collections to the state, and they were thereafter opened to the public, becoming a popular attraction in St. Petersburg. The Rumiantsev Museum was later broken up, and the minerals allotted to the Moscow State Geological Museum.

Another Russian nobleman with a passion for minerals was Count Apollos Moussin-Poushkin (1760–1805). He was primarily a chemist but in that context gathered together a substantial collection of minerals. He is known to have sold

161 Russian mineral specimens to the English collector Charles Hatchett in 1799; that same year, the British Museum purchased Hatchett's collection. A manuscript copy of the Moussin-Poushkin specimen list is still preserved there today.

Two notable Russian expatriates also assembled significant mineral collections while living in Western Europe. Count Gregorii Razumovsky (1759–1837) lived in Germany, France, and Switzerland, pursuing his geological researches which appeared in print mostly during the 1780s (*Voyages mineralogiques... 1784; Essai d'un Système des rtransitions de la Nature dans le Règne Minéral*, 1785; *Observations minérakigiques sur les environs de Vienne*, 1822; and others). He owned a collection of about 6,000 mineral specimens, mostly from Russian localities but also from Germany and Switzerland, which was said to have been built at great expense over many years. The published sale notice reads as follows:

> *This is one of the most complete collections to be found of the products of the Mineral Kingdom, containing items obtained not only from Siberia, but from the other parts of Russia, both southern and northern, as well. It includes a very interesting series of all the stones, rocks, minerals and metallic ores from that Empire, numerous specimens of a remarkable size among them. The metallic ores include samples of native gold from Olonetz on the shore of Lake Ladoga such as exist in very few collections; this mine, the workings of which are under water today, and which, moreover, have been very little studied, has not been exploited for some years. It is for this same reason that the beautiful specimens of crocoite found here are more precious today than ever. This collection fills fourteen large packing cases.*

Razumovsky sold this collection to the American collector Colonel George Gibbs, around 1800; Gibbs brought it back to Rhode Island along with other collections he had purchased in Europe, ultimately selling it to Yale University in 1825. Razumovsky was at the peak of his career at the time he sold his collection to Gibbs. Armed with fresh cash and plenty of energy, he proceeded to build a second mineral collection, which was eventually acquired, at least in part, by Austrian Prince Prosper von Sinzendorf (1733–1822).

Prince Dimitrii Golitsyn

The other Russian expatriate was Prince Dmitrii Golitsyn (1734–1803), a diplomat and naturalist who became *persona non grata* with the Czar because of his suggestion that the serfs be freed. He lived in Paris, Aachen, The Hague, Brunswick, and Jena, publishing several books on mineralogy, principally a general treatise in several editions and a glossary of mineralogy in at least two editions. He collected throughout Germany, especially in Bavaria, and received, by gift, trade, and purchase, specimens from many other countries, including his Russian homeland. Visitors were amazed at the size and variety of his collection, and the rare specimens it contained. Golitsyn was admitted to the Mineralogical Society of Jena in 1799 and was elected its president. The society provided him with a ducal castle for the storage of his enormous collection, nearly two tons in total weight; it was transferred to the University of Jena in 1905 but was destroyed by Allied bombing in 1945.

SCANDINAVIA

Mineral collecting in Copenhagen (Denmark's only major city) dates back to Ole Worm (1588–1654), a Danish physician, archeologist, and Professor of Natural Philosophy and Medicine at the University of Copenhagen. He began collecting natural history specimens (including minerals) in 1620 and devoted the first four chapters of his published catalog (*Museum Wormianum*, 1655) to the minerals in his collection. The catalog includes a fascinating engraving of his collection room, filled with stuffed animals, bones, tortoise shells, and ethnographic items, and many trays of mineral specimens are carefully organized into categories such as metals, stones, salts, and so on. His collection became famous throughout Europe, admired even by King Frederic III (1609–1670), who purchased it just prior to Worm's death, for inclusion in the Royal *Kunstkammer*. Frederic acquired other collections as well in the following years, by purchase or confiscation.

The Royal Danish *Kunstkammer* continued to be fed with *naturalia* by King Christian V (1646–1699), who published a catalog of it in 1696; King Frederic IV (1671–1730); King Fredric V (1723–1766), who published another catalog in 1759; and by King Christian VII (1749–1808), who is known to have presented mineral specimens from it to the University of Göttingen and to the King's Cabinet in Paris. In 1808, following the death of Christian VII, the mineral specimens were segregated, becoming part of the Royal Natural History Museum and later the Geological Museum of the University of Copenhagen, where some specimens still survive today.

King Christian VIII (1786–1848) built a personal mineral collection beginning at the age of seven, but it was destroyed by fire in 1794. He began forming a new one around 1800, which eventually numbered about 2,300 minerals plus assorted rocks, stones, and fossils. Following his death in 1848, it too was turned over to the University of Copenhagen.

There was no shortage of enthusiasm among the royalty of Sweden for mineral collecting. King Gustavus Adolphus (1594–1632) had begun the tradition in 1632 when he purchased an ornate cabinet (*Kunstschrank*) from Philipp Hainhofer for 6,000 gulden. It was carefully stocked with interesting specimens from the plant, animal, and mineral kingdoms. The king died that same year, and the cabinet was inherited by King Charles XI (1655–1697), who eventually donated it in 1694 to the University of Uppsala.

King Adolphus Frederick (1710–1771) of Sweden possessed an interesting mineral collection, perhaps co-owned or merely inherited by his wife, Queen Louisa Ulrica (1720–1782). The king had gone so far as to have a published catalog prepared describing his specimens (*Museum Adolpho-Fridericianum*, 1746). Louisa Ulrica, a patron of literature and science and a friend of Linnaeus, passed the collection to the University of Uppsala in 1804.

King Charles XIII (1748–1818) and his elder brother King Gustavus III (1746–1792), sons of Adolphus Frederick, also had a mineral collection, perhaps obtained in part from their parents, which was donated to the university in 1804.

King Gustavus III

King Gustavus III (1746–1792) collected minerals as well. His most elegant gift was a sumptuous cabinet, crafted in Paris in 1765 by Georg Haupt, which he filled with Swedish minerals and presented to Louis-Joseph, Prince de Condé (1736–1818) in France. It is still preserved at the Condé palace in Chantilly.

The Growth of Public MINERAL MUSEUMS

In our time, one of the greatest benefits that society can offer to the mineral collector is access to large, intelligently assembled and well-identified public mineral collections. For the student and the novice, museum collections give a first awe-inspiring view of the broad scope and fascinating variety of the inorganic crystallizations that our earth has produced. This is a tremendous advantage over having to piece together the big picture from specimens personally acquired over time, or merely reading about them in more or less unillustrated books. Before the advent of public collections, naturalists and mineral collectors commonly devoted some years of their lives to traveling about Europe visiting as many *private* collectors as possible and studying their holdings. Only this, in combination with a study of the extant literature, could yield a useful education in mineralogy.

For most of the time period preceding 1800, it must be remembered, the method of actually *analyzing* minerals in some way to identify their species was not much employed, even in universities. Minerals were identified, to a first approximation, by visual inspection alone. If they were suspected of containing precious metals they might then be submitted to an assayer; if thought to be precious gems they might be turned over to a professional lapidary. But, without an actual reference collection for study and comparison, it was almost impossible for a student or a scientist to learn how to identify minerals on his own.

Fortunately this problem was not lost on the educators or the nobility of many countries. In fact, monarchs came to realize in the 18th century that a public well informed about the appearance of valuable minerals and ores would be better able to make new discoveries of ore deposits, perhaps greatly enriching the country – and the Crown. It was also discovered by the nobility that establishing institutional collections for public betterment had the salutary effect of improving their own public image, and perhaps even making them beloved by their subjects (as happened with Empress Maria Theresa in Vienna). It was therefore in the ruler's best personal interest to provide the public with opportunities to study collections of minerals.

Consequently, as the great universities took up the teaching of natural history, they also began developing study collections. At first the instructors used specimens from their personal collections to show in class but, as time went by, professorial collections were donated to the universities, outside naturalists contributed specimens, and the aristocracy donated their own collections, accumulated through generations of royal acquisitiveness. In time, most of the major institutions where science was taught had collections of minerals, plants, and animals that were available for study by any student or visiting naturalist with a proper introduction.

Many monarchs took this concept a step further by establishing not just collections but schools and colleges devoted specifically to mining science and mineralogy. Thus were born the great mining academies in Schemnitz (1735), Potosí (1756), Kongsberg (1757), Prague (1763), Freiberg (1765), Berlin (1770), St. Petersburg (1773), Almadén (1777), and Mexico City (1792). In each case, public study collections were needed.

Mineralogists came to realize as well that their own scientific theories could be better promoted, preserved, and spread among future students and the scientific community at large if they saw to the preservation of their working collections. Some even went so far as to prepare collections of specimens or (in the cases of Romé de l'Isle and Haüy) crystal models for commercial distribution to institutions.

Private collectors also began to see that their own lifelong efforts at assembling an instructive collection of minerals would be lost if their collections were simply taken apart and scattered or (worse yet) discarded upon their deaths. The advent of permanent institutions with continuously curated collections offered the private collector a way of protecting and preserving the product of his labors long after his own death, and perhaps even immortalizing his name in some measure.

In the 20th century, wealthy individuals, business people, and corporations discovered that fostering public collections helped put their city "on the map" with the growing community of scholars and private collectors, especially following the big boom in "rockhounding" interest that arose after World War II. The public had gained a new awareness of minerals through rockhounding and lapidary activities, and support of museums and public mineral collections yielded public relations benefits. And, of course, numerous wealthy individuals and even a few corporations became enthralled with minerals and built significant collections of their own, which were ultimately acquired by museums, thus preserving their names for history in a way that mere business success could not.

So it has been in everyone's interest that institutional mineral collections should be established, nurtured, contributed to and made accessible to the public.

CONCLUSIONS

The passion for collecting minerals focused mineralogical studies in three major areas: the chemical analysis of minerals, the structure and physical properties of mineral crystals, and the rational classification of mineral substances. The progressive acquisition of knowledge in these areas jointly contributed to the founding of the science of mineralogy (Greene and Burke, 1978).

The beginning of mineral collecting in the 16th century was therefore a necessary prerequisite to the development of the first systematic classification schemes for minerals. The critical importance of amateur mineralogists and aristocratic collectors can hardly be overemphasized. Peter Embrey and Robert Symes (1987) of the British Museum have made the same observation:

> *Mineralogy as a science, in common with other branches of natural history, owes an inestimable debt to the enthusiasm of amateur collectors; for it is nothing if not specimen-based, and our public museums would be poor things indeed had they not first been based on private collections. Much of their growth then has been due to generous gifts or bequests, and to the purchase of whole or selected parts of private collections.*

It is interesting to note that commerce in mineral specimens was already well developed in the context of pharmaceuticals, and that specimen dealing contributed significantly to the very first collections, which gave birth to the science of mineralogy. It is also worth emphasizing that the practice among German miners of collecting mineral specimens as keepsakes and heirlooms is older than science itself, and no doubt contributed greatly to those first historic collections that proved so important for the development of mineralogy and mining technology.

Perhaps the most striking observation that can be made from a study of the origins of mineral collecting is how little the basic process has changed. The first collectors field-collected, traded, and purchased their specimens, took pride in them, showed them off to visiting like-minded enthusiasts, and stored them in cabinets and cases not significantly different from those used in all later centuries right up to the present day. The collections were studied by mineralogists and new species described beginning with Agricola himself.

Perhaps equally striking is the alarming vulnerability of mineral collections to the ravages of time. Illustrations of specimens from early collections, rendered on fragile paper, routinely outlive the specimens themselves. Countless thousands of fine mineral specimens that were lovingly collected, studied, and preserved by early collectors are now lost, badly damaged, or destroyed. The preservation and conservation of mineral specimens and their associated historical provenance is an uphill battle that every collector and especially every museum curator must fight. And, as in the past, the publication of specimen illustrations will provide some insurance for future generations in case the original specimens are later lost or damaged.

The avocation and profession of mineralogy continue unabated in the 21th century. Despite over 400 years of accumulation and study, people are still finding new occurrences and new species of a beauty and fascination to rival anything in the past. What is different, and what separates modern collectors from their early counterparts, is the vast mass of knowledge and literature built up over that time, which is today available to any mineral collector. Thousands of books and numerous journals have been published and are easily accessible. Millions of specimens are preserved for study in museums, and millions more are in private collections or available for purchase from dealer stocks. Tens of thousands of specimens have been illustrated in the literature and thus may be studied by collectors worldwide. The study of the history of mineral collecting serves to remind modern collectors not only of their age-old traditions and their brotherhood with science, but of how fortunate people are in modern times to have so many resources that were unavailable to the pioneers of mineral collecting.

SELECTED BIBLIOGRAPHY

BANCROFT, P. *Gem & Crystal Treasures.* Tucson: Western Enterprises and Mineralogical Record, 1984.

BANCROFT, P. B. *The World's Finest Minerals and Crystals.* New York: Viking Press, 1973.

BARLOW, F. J., R. W. JONES, and G. L. LABERGE. *The F. John Barlow Mineral Collection.* Appleton, Wisconsin: Sanco Publishing, 1996.

BLAUWET, D., B. SMITH, and C. SMITH. "A guide to the mineral localities of the Northern Areas, Pakistan." *Mineralogical Record* 28 (1997): 183–200.

BURLAKOV, E. V. "The Puiva deposit, Subpolar Urals, Russia." *Mineralogical Record* 30 (1990): 451–65.

CASSEDANNE, J. P. "Famous mineral localities: The Córrego Frio mine and vicinity, Minas Gerais, Brazil." *Mineralogical Record* 14 (1983): 227–37.

———. "Famous mineral localities: The Ouro Preto topaz mines." *Mineralogical Record* 20 (1989): 221–33.

CASSEDANNE, J. P. and J. LOWELL. "Famous mineral localities: The Virgem da Lapa pegmatites." *Mineralogical Record* 13 (1982): 19–28.

COOK, R. B. "Famous mineral localities: Chañarcillo, Chile." *Mineralogical Record* 10 (1979): 197–204.

CROWLEY, J. A., R. H. CURRIER, and T. SZENICS. "Mines and minerals of Peru." *Mineralogical Record* 28 (1997): 1–98.

CROWLEY, J. A. and N. A. RADFORD. "Pyromorphite from the Couer d'Alene district, Idaho." *Mineralogical Record* 13 (1982): 273–85.

CURRIER, R. H. "Amethyst, agate and calcite 'skunks' from Uruguay." *Rocks & Minerals* 72 (1997): 94–105.

DUNN, P. J. "Rare minerals of the Kombat mine, Namibia." *Mineralogical Record* 22 (1991): 421–25.

DUNN, P. J., C. W. WOLFE, P. B. LEAVENS, and W. E. WILSON. "Hydroxyl-herderite from Brazil and a guide to species nomenclature for the herderite/hydroxyl-herderite series." *Mineralogical Record* 10 (1979), 5–11.

EMBREY, P. G. and R. F. SYMES. *Minerals of Cornwall and Devon.* London: British Museum (Natural History) and Tucson: Mineralogical Record, 1987.

FEININGER, T. "Emerald mining in Colombia: history and geology." *Mineralogical Record* 1 (1970): 142–49.

FISHER, J. "Gem and rare-element pegmatites of southern California." *Mineralogical Record* 33 (2002): 363–407.

GAINES, R. V. "Interview: Allan Caplan." *Mineralogical Record* 11 (1980): 351–60.

GAINES, R. V. and D. THADEU. "The minerals of Panasqueira, Portugal." *Mineralogical Record* 2 (1971): 73–78.

GAUTIER, G., A. FRANÇOIS, M. DELIENS, and P. PIRET. "Famous mineral localities: The uranium deposits of the Shaba region, Zaire." *Mineralogical Record* 20 (1989), 265–88.

GEBHARD, G. *Tsumeb, A Unique Mineral Locality.* Grossenseifen, Germany: GG Publishing, 1999.

———. *Tsumeb, eine deutsch-afrikanische Geschichte*. Verlag Christel Gebhard-Giesen, 1991.

GRAEME, R. W. "Famous mineral localities: Bisbee, Arizona." *Mineralogical Record* 12 (1981): 258–319.

GRANT, R. and W. E. WILSON. "Famous mineral localities: Dalnegorsk, Primorskiy Kray, Russia." *Mineralogical Record* 32 (2001): 3–30.

GREENE, J. C. and BURKE, J. G. "The science of minerals in the age of Jefferson," *Transactions of the American Philosophical Society*, 68, part 4 (1978), 1–113.

GRICE, J. D. and R. WILLIAMS. "Famous mineral localities: The Jeffrery mine, Asbestos, Quebec." *Mineralogical Record* 10 (1979): 69–80.

GÜBELIN, E. *Die Edelsteine der Insel Ceylon*. Gubelin Lucerne, 1968.

GUNAWARDENE, M. and M. S. RUPASINGHE. "The Elahera gem field in central Sri Lanka." *Gems & Gemology* 22, no. 2 (1986), 80-95.

HUGHES, R. W. *Ruby & Sapphire*. Boulder, Colorado: RWH Publishing, 1997.

JENSEN, M. "The Meickle mine, Elko County, Nevada." *Mineralogical Record* 30 (1999), 187–96.

JOHNSEN, O. "Famous mineral localities: the Kongsberg silver mines, Norway." *Mineralogical Record* 17 (1986), 19–36.

KAMPF, A. R. and P. C. KELLER. "The Colorado Quartz mine, Mariposa County, California: a modern source of crystallized gold." *Mineralogical Record* 13 (1982), 347–54.

KASMI, A. H., J. J. PETERS, and H. OBODDA. "Gem pegmatites of the Shingus-Dusso area, Gilgit, Pakistan." *Mineralogical Record* 16 (1985), 393–411.

KEARNS, L. E. and H. CAMPBELL, III. "Famous mineral localities: The Elmwood and Gordonsville zinc mines." *Mineralogical Record* 9 (1978), 213–18.

KELLER, P. C. "The Capão topaz deposit, Ouro Preto, Minas Gerais, Brazil." *Gems & Gemology* 19, no. 1 (1983), 12–20.

———. "The rubies of Burma: A review of the Mogok Stone Tract." *Gems & Gemology* 19, no. 4 (1983), 209–19.

KIEVLENKO, E. YA. *Geology of Gems*. Moscow: Ocean Pictures Ltd., 2003.

LANCASTER, K. "Crocoite and its increasing scarcity." *Mineralogical Record* 8 (1977), 24–26.

LEICHT, W. "California gold." *Mineralogical Record* 13 (1982), 375–87.

LUCIO, A. and R. V. GAINES. "The minerals of the Morro Velho gold mine." *Mineralogical Record* 4 (1973), 224–29.

MAKKI, M. F. "Spectacular new apophyllite from India's Deccan Trap Basalts." *Rocks & Minerals* 77 (2002), 126–29.

MOORE, T. P. A collection of diamond crystals, with notes on the science, history and worldwide localities of diamonds. *Mineralogical Record* 35 (2004), 9–30, 53–54, 63.

MOORE, T. P., B. K. LEES, K. J. WENRICH, S. VOYNICK, J. A. MURPHY, J. F. HURLBUT, T. J. REYNOLDS, R. AUMENTE-MODRESKI, D. MISANTONI, and ML. L. SILBERMAN, "The Sweet Home mine, Park County, Colorado." *Mineralogical Record* 29, no. 4 (1998), 1–153.

MOORE, T. P. and P. K. M. MEGAW. "Famous mineral localities: The Ojuela mine, Mapimi, Durango, Mexico." *Mineralogical Record* 34, no. 5 (2003), 1–91.

OLSON, D. K. "Native silver occurrences in the copper mines of Upper Michigan." *Mineralogical Record* 17 (1986), 37–48.

OTTENS, C. "Minerals of the Deccan Traps, India." *Mineralogical Record* 34 (2003), 1–82.

PANCZNER, W. D. *Minerals of Mexico.* New York: Van Nostrand Reinhold Company, 1987.

PAPP, G., S. SZAKALL, and T. WEISZBERG. "A brief history of the collections of rocks and minerals in Hungary." In *Annals of the History of Hungarian Geology, special issue 3: Museums and Collections in the History of Mineralogy, Geology and Paleontology in Hungary*. Edited by G. Vitalis and T. Kecskemeki. Budapest: the Hungarian Geological Survey and the Hungarian Geological Society, 1991.

POPOVA, V. I., V. A. POPOV, and A. A. KANONEROV. "Murzinka: Alabashka pegmatite field." *Mineralogical Almanac* 5 (2002), 1–128.

PROCTOR, K. "Gem pegmatites of Minas Gerais, Brazil: Exploration, occurrence, and aquamarine deposits." *Gems & Gemology* 20 (1985b), 78–100.

———. "Gem pegmatites of Minas Gerais, Brazil: The tourmalines of the Araçuai districts." *Gems & Gemology* 21, no. 1 (1985a), 3–19.

———. "Gem pegmatites of Minas Gerais, Brazil: The tourmalines of the Governador Valadares district." *Gems & Gemology* 21, no. 2 (1985b), 86–104.

RADFORD, N. and J. A. CROWLEY. "The Bunker Hill mine, Kellogg, Shoshone County, Idaho." *Mineralogical Record* 12 (1981), 339–47.

ROE, A. "Washington A. Roebling, his life and his mineral collection." *Mineralogical Record* 21 (1990), 13–30.

SAUER, D. A., A. S. KELLER, and S. F. MCCLURE. "An update on Imperial topaz from the Capão mine, Minas Gerais, Brazil." *Gems & Gemology* 32, no. 4 (1996), 232–41.

SCHULZ, E. "Notes on the history of collecting and of museums, in the light of selected literature of the sixteenth to the eighteenth century." *Journal of the History of Collections*, 2 (1990), 205–218.

SEEMAN, R. "Famous mineral localities: Knappenwand, Untersulzbachtal, Austria." *Mineralogical Record* 17 (1986), 167–81.

SINKANKAS, J. *Emerald and Other Beryls.* Radnor, Pennsylvania: Chilton Book Company, 1981.

SWOBODA, E. R. "History of the Tourmaline Queen mine, San Diego County, California." *Mineralogical Record* 33 (2002), 409–25.

WILSON, M. L. and S. J. DYL, III. "The Michigan Copper Country." *Mineralogical Record* 23, no. 2 (1992), 1–76.

WILSON, W. E. "Connoisseurship in minerals." *Mineralogical Record* 21 (1990), 7-12.

———. "Famous mineral localities: The Erupcion/Ahumada mine, Los Lamentos district, Chihuahua, Mexico." *Mineralogical Record* 34, no. 6 (2003), 4–31.

———. "Famous mineral localities: Lavra Berilo Branco, the original 'Sapucaia' rose quartz occurrence, Minas Gerais, Brazil." *Mineralogical Record*, 30 (1999) 361–65.

———. "The history of mineral collecting, 1530–1799." *Mineralogical Record* 25 (1994), 1–243.

———, ed. "Tsumeb, the world's greatest mineral locality." *Mineralogical Record* 8, no. 3 (1977), 8–111.

WILSON, W. E. and J. A. BARTSCH. "Special supplement: Minerals of the Houston Museum of Natural Science." *Mineralogical Record* 23 (1992), 1–34 (separately paginated insert).

WILSON, W. E. and P. J. DUNN. "Famous mineral localities: The Kalahari Manganese Field." *Mineralogical Record* 9 (1978), 137–53.

WILSON, W. E., C. L. JOHNSTON, and E. R. SWOBODA. "Jeremejevite from Namibia." *Mineralogical Record* 33 (2002), 289–301.

WILSON, W. E. and A. PETROV. "Famous mineral localities: Cerro Rico de Potosí, Bolivia." *Mineralogical Record* 30 (1999), 9–36.

PHOTOGRAPHY CREDITS

All mineral photography by Jeffrey Scovil, Tucson, except the following plates: 3, Azurite; 20 and 21, Cerussite; 22, Chrysoberyl; 25, Corundum; 26, Corundum; 29, Diamond; 31, Elbaite; 40, Fluorite; 52, Mesolite; 57, Proustite; 62, Rhodochrosite; 74, Spodumene; 75, Sulfur; 78, Torbernite; and 79, Zoisite, by Erica and Harold Van Pelt, Hollywood.

Historical photography from "The Paths of Discovery" reprinted with permission from *Mineralogical Record*/Western Publishing: 7, Chivor, Peter Bancroft photo; 25, Mogok, Edward Gubelin photo; 29, Musonoi, courtesy Union Miniere Co.; 75, Agrigento, Pier Bruno Scortecci photo. Reprinted with permission from *Mineralogical Record*: 2 and 50, Ojuela, courtesy Mapimi Mining Museum; 5 and 73, Bisbee, courtesy Bisbee Mining & Historical Museum; 23, Keewenaw, Michigan Technological Univ. Archives; 32, Salinas, Francisco Müller Bastos photo; 39, Dalnegorsk, Ray Grant Photo; 43, Jeffrey, Marco Amabili photo; 47, Mile72, Chris Johnston photo; 58, Bunker Hill, Norm Radford photo; 62, Sweet Home, Collector's Edge photo; 68, Kongsberg, courtesy Kongsberg Mining Museum; 74, Pala Chief, Brendan Laurs photo. Reprinted with permission from Mineralogical Record Library and Photo Archives: 1, Arizpe; 56, Potosi; 70, Quincy; 3, 20, 49, 51, 55 and 71, Tsumeb, courtesy Ilse Schatz photo collection; 45, Xanda, Jack Lowell photo.

Painting of John Stuart, the Earl of Bute, reprinted in "A Royal Passion" courtesy of the National Portrait Gallery, London.